Senior

Biology 1

Student Workbook

Model Answers: 2011

This model answer booklet is a companion publication to provide answers for the exercises in the Senior Biology 1 Student Workbook 2011 edition. These answers have been produced as a separate publication to keep the cost of the workbook itself to a minimum, as well as to prevent easy access to the answers by students. In most cases, simply the answer is given with no working or calculations described. A few, however, have been provided with greater detail because of their more difficult nature.

www.thebiozone.com

ISBN 978-1-877462-61-0

Published by BIOZONE International Ltd

Additional copies of this Model Answers book may be purchased directly from the publisher

NORTH & SOUTH AMERICA, AFRICA:
BIOZONE International Ltd.
P.O. Box 13-034, Hamilton 3251, **New Zealand**
Telephone: +64 7-856-8104
FREE Fax: 1-800-717-8751 (USA-Canada)
FREE Phone: 1-866-556-2710 (USA-Canada)
FAX: +64 7-856-9243
E-mail: sales@biozone.co.nz

EUROPE & MIDDLE EAST:
BIOZONE Learning Media (UK) Ltd.
Bretby Business Park, Ashby Road, Bretby, Burton upon Trent, DE15 0YZ, **UK**
Telephone: +44 1283-553-257
FAX: +44 1283-553-258
E-mail: sales@biozone.co.uk

ASIA & AUSTRALIA:
BIOZONE Learning Media Australia
P.O. Box 2841, Burleigh BC, QLD 4220, **Australia**
Telephone: +61 7-5535-4896
FAX: +61 7-5508-2432
E-mail: sales@biozone.com.au

Contents

Contents

Contents

Investigations in Field Science (page 12)

1. The major features of the habitat were given the numbers 1-5 so that a numerical scale was created based on the degree of disturbance in the habitat.
2. Only cover types taking up more than 20% of the habitat were used as descriptors.
3. Fish were trapped over a long period of time in many areas that covered more than the known range of mudfish. In each area, traps were set at the same time of the day over a wide area, in a uniform way.
4. Mudfish weight and length were measured.
5. Mudfish prefer minimally disturbed environment.
6. Because few areas were found to have a disturbance rating of 1 or 2 the preferences there were too unreliable to distinguish between them.

Hypotheses and Predictions (page 13)

1. Prediction: Woodlice are more likely to be found in moist habitats than in dry habitats.
2. (a) **Bacterial cultures**:
   Prediction: Bacterial strain A will grow more rapidly at 37°C than at room temperature (19°C).
   Outline of the investigation: Set up agar plates of bacterial strain A, using the streak plating method. Place 4 plates in a 37°C incubator and 4 on the lab bench. Leave all 8 plates for the same length of time (e.g. 24 hours), with all other conditions identical. Measure the coverage of the agar plates with bacteria (as a percentage).
   (b) **Plant cloning**:
   Prediction: A greater concentration of hormone A increases the rate of root growth in plant A.
   Outline of the investigation: Set up 6 agar plates infused with increasing concentrations of hormone A (e.g. 1 mgl^{-1}, 5 mgl^{-1}, 10 mgl^{-1}, 50 mgl^{-1}, 100 mgl^{-1}, 500 mgl^{-1}), and each plate with 12 clones of plant A. Measure root length each day for 20 days.

Planning an Investigation (page 15)

1. Aim: To investigate the effect of temperature on the rate of catalase activity.
2. Hypothesis: The rate of catalase activity is dependent on temperature.
3. (a) Independent variable: Temperature.
   (b) Values: 10-60°C in uneven steps: 10°C, 20°C, 30°C, 60°C.
   (c) Unit: °C
   (d) Equipment: A means to maintain the test-tubes at the set temperatures, e.g. water baths; equilibrate all reactants to the required temperatures in each case, before adding enzyme to the reaction tubes.
4. (a) Dependent variable: Height of oxygen bubbles.
   (b) Unit: mm
   (c) Equipment: Ruler; place vertically alongside the tube and read off the height (directly facing as you would a meniscus).
5. (a) Each temperature represents a treatment.
   (b) No. of tubes at each temperature = 2
   (c) Sample size: for each treatment = 2
   (d) Times the investigation repeated = 3
6. It would have been desirable to have had an extra tube with no enzyme to determine whether or not any oxygen was produced in the absence of enzyme.
7. Variables that might have been controlled (a-c):
   (a) Catalase from the same batch source and with the same storage history. Likewise for the H_2O_2. Storage and batch history can be determined.
   (b) Equipment of the same type and size (i.e. using test-tubes of the same dimensions, as well as volume). This could be checked before starting.
   (c) Same person doing the measurements of height each time. This should be decided beforehand.

   Note that some variables were controlled: the test-tube volume, and the volume of each reactant. Control of measurement error is probably the most important after these considerations.
8. Controlled variables should be monitored carefully to ensure that the only variable that changes between treatments (apart from the biological response variable) is the independent (manipulated) variable.

Experimental Method (page 17)

1. Increasing the sample size is the best way to take account of natural variability. In the example described, this would be increasing the number of plants per treatment. **Note**: Repeating the entire experiment as separate trials (as described) is a compromise, usually necessitated by a lack of equipment and other resources. It is not as good as increasing the sample size in one experiment run at the same time, but it is better than just the single run of a small sample size.
2. If all possible variables except the one of interest are kept constant, then you can be more sure that any changes you observe in your experiment (i.e. differences between experimental treatments) are just the result of changes in the variable of interest.
3. Only single plants were grown in each pot to exclude the confounding effects of competition between plants (this would occur if plants were grown together).
4. Physical layout can affect the outcome of experimental treatments, especially those involving growth responses in plants. For example, the physical conditions might vary considerably with different placements along a lab bench (near the window vs central). Arranging treatments to minimize these effects is desirable.

   Checklist to be completed by the student.

Recording Results (page 19)

1. See the results table at the top of the next page.
2. The table would be three times as big in the vertical dimension; the layout of the top of the table would be unchanged. The increased vertical height of the table would accommodate the different ranges of the independent variable (full light, as in question 1, but also half light, and low light. These ranges would have measured values attached to them (they should be quantified, rather than subjective values).

		Trial 1 $[CO_2]$ in ppm (day 0)											Trial 2 $[CO_2]$ in ppm (day 2)											Trial 3 $[CO_2]$ in ppm (day 4)										
		Minutes											Minutes											Minutes										
	Set up no.	0	1	2	3	4	5	6	7	8	9	10	0	1	2	3	4	5	6	7	8	9	10	0	1	2	3	4	5	6	7	8	9	10
Full light conditions	1																																	
	2																																	
	3																																	
	Av.																																	

Variables and Data (page 20)

1. (a) Leaf shape: **Qualitative**
   (b) Number per litter: **Quantitative**, discontinuous
   (c) Fish length: **Quantitative**, continuous.
2. Quantitative data are more easily and meaningfully analysed, using descriptive and inferential statistics.
3. Measure wavelength (in nm) with a spectrophotometer; which measures light intensity as a function of the color, or more specifically, the wavelength of light.
4. (a) Many variables could be chosen, e.g. viability (dead or alive), gender, species, presence or absence of a feature, flower color. These data are categorical; no numerical value can be assigned to them.
   (b) These data are semi-quantitative because arbitrary numerical values have been assigned to a qualitative scale. The numbers are correct in a relative sense, but do not necessarily indicate the true quantitative values.

Manipulating Raw Data (page 21)

1. Basic transformations are performed on raw data so that important features (e.g. trends) of the data can be easily identified.
2. (a) Percentage.
   (b) Calculation of the percentage of plant species enables a comparison of the relative number of species at each habitat.
3. Incidence of cyanogenic clover in different areas

Clover plant type	Frost free area		Frost prone area		Totals
	Number	%	Number	%	
Cyanogenic	120	76	22	**15**	**142**
Acyanogenic	38	**24**	120	**85**	**158**
Total	158	**100**	**142**	**100**	**300**

Constructing Tables (page 22)

1. (a)-(b) any two of the following:
   - Tables provide a systematic record of information.
   - Tables provide a way of condensing data.
   - Tables provide a summary of results.
   - Tables show trends and relationships in the data.
2. Data might first be tabulated so that the researcher can easily identify any trends in the data. This will enable them to decide the best method of graphing the data.
3. (a) Presentation of descriptive statistics in a table allows the data to be summarised so that trends or patterns can be identified more easily than by the presentation of raw data alone. It also aids the researcher in deciding the best way to graph the data to show the trend.
   (b) Inclusion of the measure of spread allows the researcher to determine if the data is normally distributed or skewed. This will determine how the data will be further manipulated.
4. The control values should be placed at the top of the table so that it is easily identified and can be easily compared with the values obtained for the treatments. This allows the researcher and readers to determine if the treatments are having any real effect.

Constructing Graphs (page 23)

1. Graphs visually show a trend or relationship in data using a minimum of space.
2. (a) An appropriate scale must be used to show the trend in the data to its best effect. For example, if the scale is too compressed then it is difficult to see the trend or pattern.
   (b) A floating axis may be used when there is a large gap between zero and where the data begin. This allows for trends to be more easily visualised.
3. (a) The time intervals on the X-axis are evenly spaced even though the time increments at which the data were recorded are not.
   (b) If the graph were plotted correctly then the data points would be stretched out. When the line of best fit was applied the slope of the plotted data would be much flatter than the original.

Drawing Bar Graphs (page 24)

1. (a) See table below:

Species	Site 1	Site 2
Ornate limpet	21	30
Radiate limpet	6	34
Limpet sp. A	38	-
Limpet sp. B	57	39
Limpet sp. C	-	2
Catseye	6	2
Topshell	2	4
Chiton	1	3

   (b) Bar graph: See page 4 for graph solutions.

Drawing Histograms (page 25)

1. (a) Tally chart totals as below:

Weight group	Total
45-49.9	1
50-54.9	2
55-59.9	7
60-64.9	13
65-69.9	15
70-74.9	13
75-79.9	11
80-84.9	16
85-89.9	9
90-94.9	5
95-99.9	2
100-104.9	0
105-109.9	1

(b) Histogram: See next page of graph solutions.

Drawing Pie Graphs (page 26)

1. (a) Tabulated data:

Food item in diet	Ferrets		Rats		Cats	
	% in diet	Angle (°)	% in diet	Angle (°)	% in diet	Angle (°)
Birds	23.6	85	1.4	5	6.9	25
Crickets	15.3	55	23.6	85	-	-
Insects	15.3	55	20.8	75	1.9	7
Voles	9.2	33	-	-	19.4	70
Rabbits	8.3	30	-	-	18.1	65
Rats	6.1	22	-	-	43.1	155
Mice	13.9	50	-	-	10.6	38
Fruits	-	-	40.3	145	-	-
Leaves	-	-	13.9	50	-	-
Unid.	8.3	30	-	-	-	-

(b) Pie graphs: See next page of graph solutions.

Drawing Kite Graphs (page 27)

1. (a) Table:

Distance from mouth (km)	Wet weight (g m^{-2})		
	Stm A	Stm B	Stm C
0	0.4	0.4	0
0.5	0.5	0.6	0.5
1.0	0.4	0.1	0
1.5	0.3	0.5	0.2
2.0	0.3	0.4	-
2.5	0.6	0.3	-
3.0	0.1	-	-
3.5	0.7	-	-
4.0	0.2	-	-
4.5	2.5	-	-
5.0	0.3	-	-

(b) Kite graph: See next page of graph solutions.

Drawing Line Graphs (page 28)

1. (a)

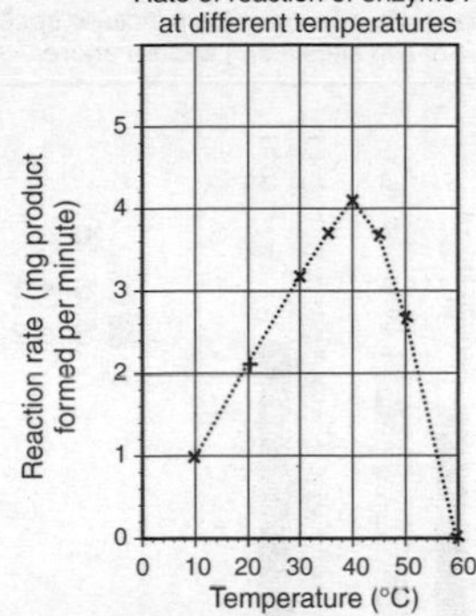

(b) Rate of reaction at 15°C = 1.6 mg product min^{-1}

2. (a) Line graph: See next page of graph solutions.
   (b) The data suggest that the deer population is being controlled by the wolves. Deer numbers increase to a peak when wolf numbers are at their lowest; the deer population then declines (and continues declining) when wolf numbers increase and then peak. **Note**: A scenario of apparent control of the deer population by the wolves is suggested, but not confirmed, by the data. In natural systems, this suggestion (of prey control by a large predator) *may* be specious; most large predators do not control their prey (except perhaps at low population densities in certain systems), but are themselves controlled by the numbers of available prey, which are regulated by other factors such as food availability. In this case, the wolves were introduced for the purpose of controlling deer and were probably doing so. However, an equally valid interpretation of the data could be that the wolves are responding to changes in deer numbers (with the usual lag inherent in population responses), and the deer were already peaking in response to factors about which we have no information.

3. (a) Line graph and (b) point at which shags and nests were removed: See the top of page 5.

Interpreting Line Graphs (page 31)

1. (b) **Slope**: Negative linear relationship, with constantly falling slope.
   Interpretation: Variable Y decreases steadily with increase in variable X.
   (c) **Slope**: Constant, with slope = 0.
   Interpretation: Increase in variable X does not affect variable Y.
   (d) **Slope**: Slope rises and then becomes 0.
   Interpretation: Variable Y initially increases with increase in variable X, then levels out (no further increase with increase in variable X).
   (e) **Slope**: Rises, peaks, and then falls (parabolic).
   Interpretation: Variable Y initially increases with increase in variable X, peaks and then declines with further increase in variable X.
   (f) **Slope**: Exponentially increasing slope.
   Interpretation: As variable X increases, variable Y increases exponentially.

Drawing bar graphs:

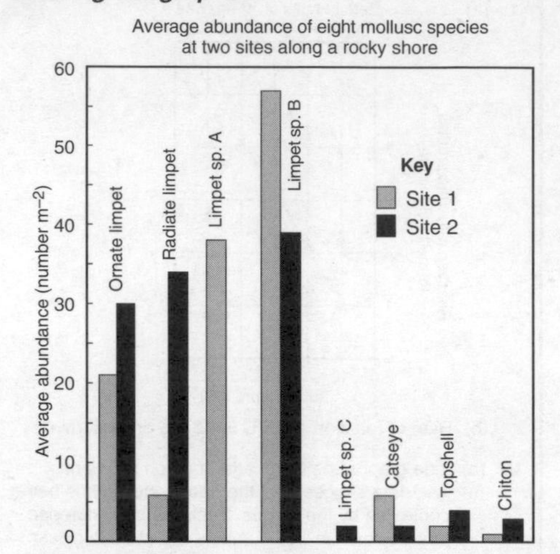

Drawing histograms:

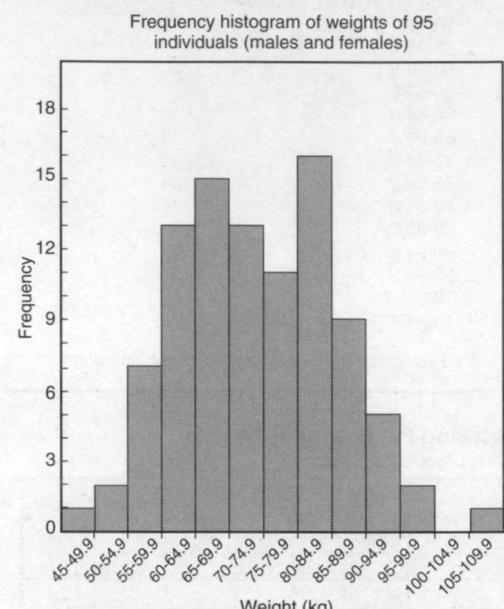

Drawing pie graphs:

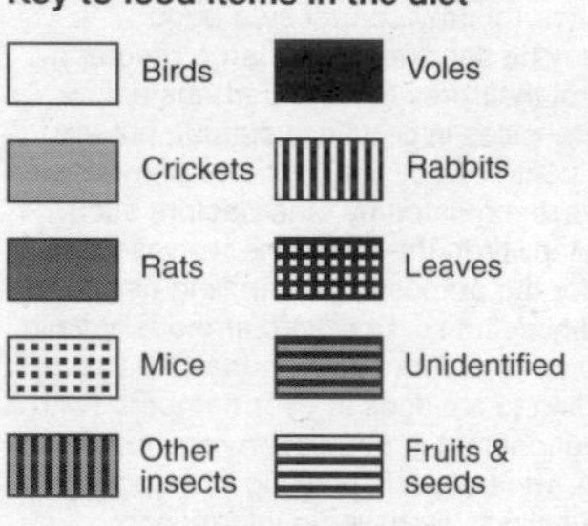

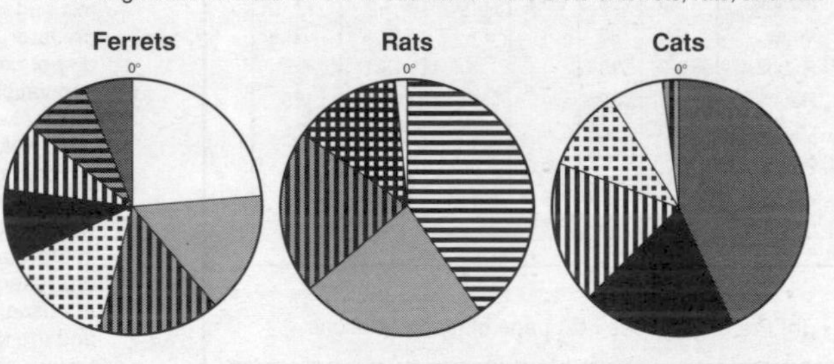

Drawing kite graphs:

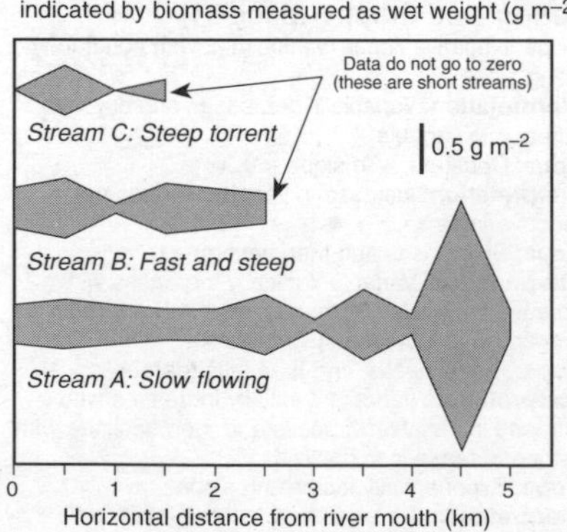

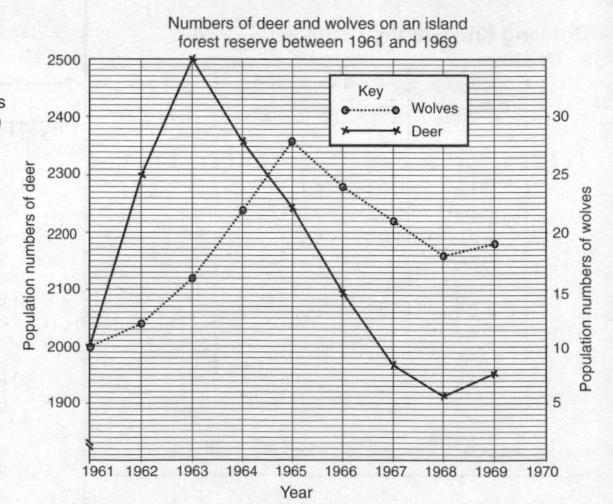

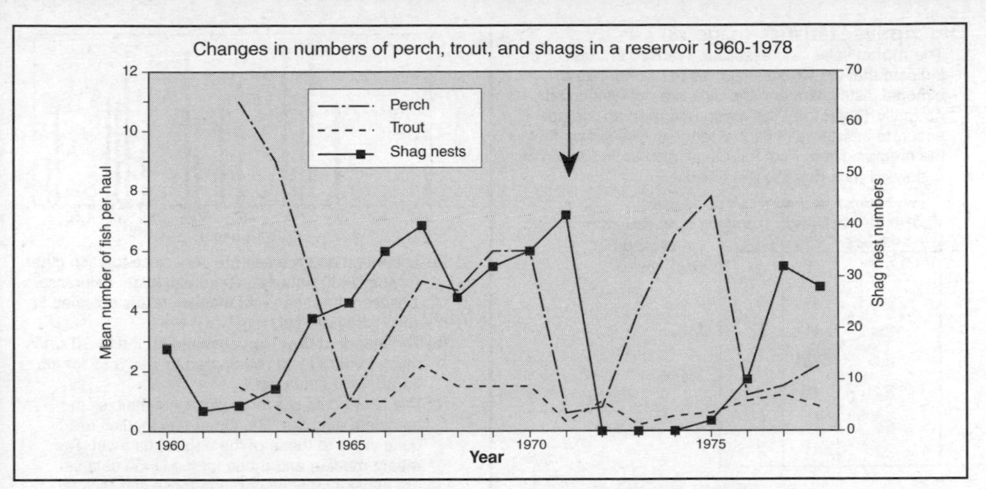

2. (a) Perch population fluctuations follow shag population fluctuations closely.
   (b) The evidence suggests that the fluctuations of shag and trout numbers are not related as the height of trout fluctuations in 1967 is reached before that of shag numbers.

Drawing Scatter Plots (page 32)

1. Scatter plot and fitted curve:

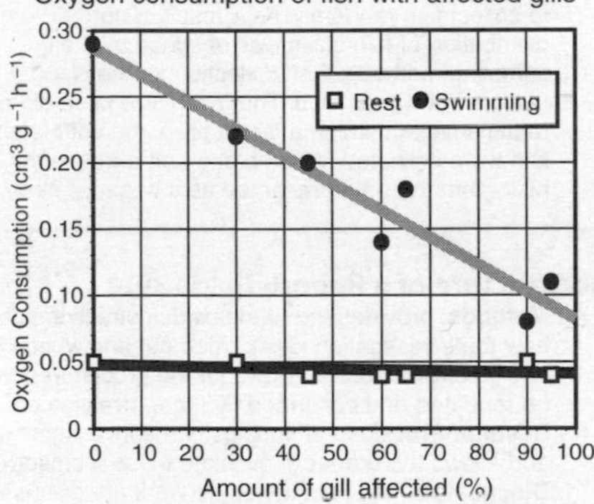

2. (a) At rest: No clear relationship; the line on the graph appears to have no significant slope (although this could be tested). **Note**: there is a slight tendency for oxygen consumption to fall as more of the gill becomes affected, but the scatter of points precludes making any conclusions about this.
   (b) Swimming: A negative linear relationship; the greater the proportion of affected gill, the lower the oxygen consumption.

3. The gill disease appears to have little or no effect on the oxygen uptake in resting fish.

Biological Drawings (page 33)

1. (a)-(h). Eight of the following, in any order:
   - Lines cross over each other and are angled.
   - Cells are inaccurately drawn: they are not closed shapes, they do not even nearly represent what is actually there; there are overlaps.
   - There is no magnification given.
   - The drawing is cramped at the top corner.
   - Labels are drawn on an angle.
   - There is no indication of whether the section is a cross section or longitudinal section.
   - There is a line to a cell type that has no label
   - Shading is inappropriate and unnecessary; it does not indicate anything.
   - The material being drawn has not been identified accurately in the title by species.

2. Student's response required here. Some desirable features are shown in the figure on the top of the next page, but page position and size cannot be shown.

3. A **biological drawing** is designed to convey useful information about the structure of an organism. From such diagrams another person should be able to clearly identify similar organisms and structures. By contrast, **artistic drawings** exhibit 'artistic license' where the image is a single person's impression of what they saw. It may not be a reliable source of visual information about the structure of the organism.

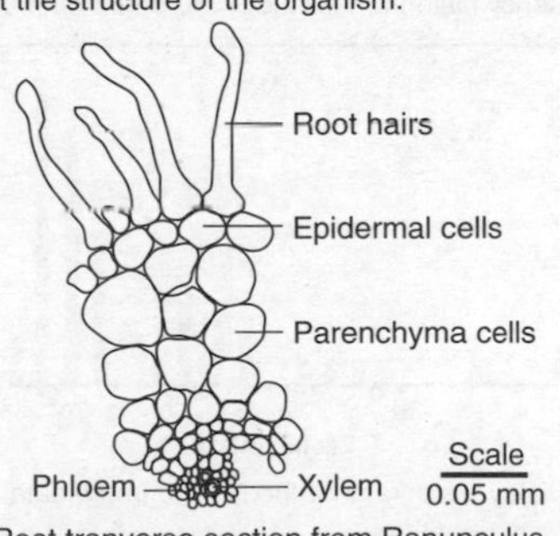

Root tranverse section from Ranunculus

Descriptive Statistics (page 35)

1. The **modal** value and associated ranked entries indicate that the variable (spores per frond) has a bimodal distribution and the data are not distributed normally. (Therefore) the **mean** and **median** are not accurate indicators of central tendency. Note also that the median differs from the mean; also an indication of a skewed (non-normal) distribution.

2.

Beetle mass (g)	Tally	Total
2.1	I	1
2.2	II	2
2.4	II	2
2.5	IIII	4
2.6	III	3
2.7	I	1
2.8	II	2

Median = 8th value when in rank order = 2.5

Mode = 2.5

Mean = 2.49 ~ 2.5

Interpreting Sample Variability (page 37)

1. (a) 496/689 values within ± 1sd of the mean = 72% (48±7.8, i.e. between 40.2 and 55.8)
   (b) 671//689 values within ± 2 sd of the mean = 97% (48± 15.6, i.e. between 32.4 and 63.6)
   (c) The data are very close to being normally distributed about the mean (normal distribution + 67% of values lie within 1sd of the mean and 95% of values lie between 2 sd of the mean).

2. The mean and the median are very close.

3. N = 30 data set
   (a) **Mean** = 49.23
   (b) **Median** = 49.5
   (c) **Mode** = 38
   (d) **Sample variance** =129.22
   (e) **Standard deviation** = 11.37

4. N = 50 data set
   (a) **Mean** = 61.44
   (b) **Median** = 63
   (c) **Mode** = 64
   (d) **Sample variance** = 14.59
   (e) **Standard deviation** = 3.82

5. Frequency histogram for the N=50 perch data set.

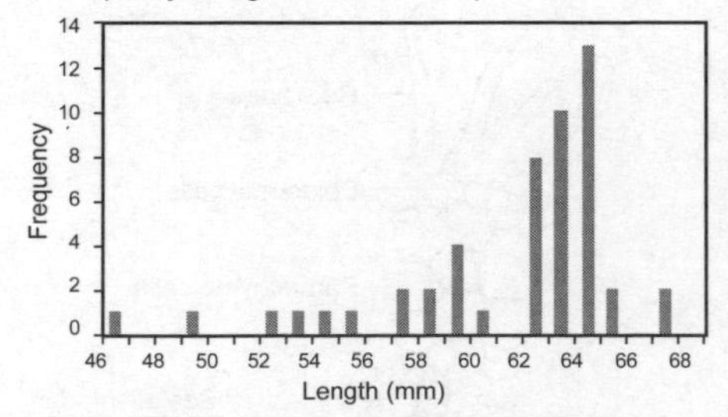

Frequency histogram for the N = 30 perch data set.

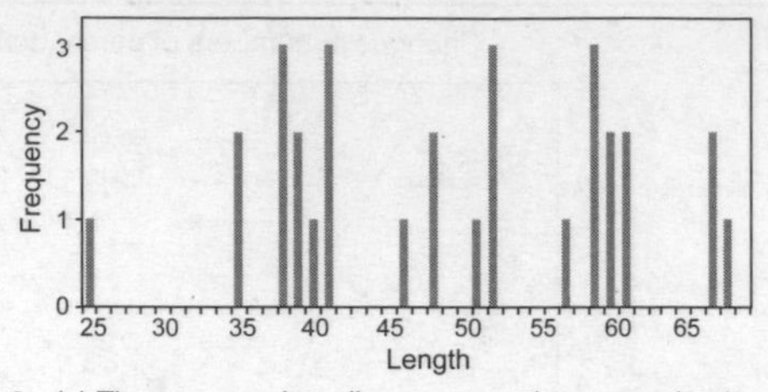

6. (a) The mean and median are very close to each other for the N=30 data set. There is a larger difference between the mean and median values obtained in the N=50 data set.
   (b) The standard deviation obtained for the N=30 set is much larger (11.37) compared to only 3.82 for the larger N=50 data set.
   (c) The N=30 data set more closely resembles the complete data set. The mean and median are quite close to those of the original data set. The mean, median and mode for the N=50 data set are considerably higher than those statistics for the complete data set. The sample variance and standard deviation values for the complete data set fall between those of the two smaller data sets.

7. (a) The frequency histogram for the N=30 data set shows a relatively normal distribution of data. The frequency histogram for the N=50 data set shows a non-normal distribution which is skewed to the right (negative skew).
   (b) The person who collected the sample in the N=30 data set used equipment and techniques designed to collect fish randomly. As a result, a normal distribution of fish sizes was obtained by their sampling methods. Fish collection for the N=50 sample set was biased. The mesh size used did not retain smaller fish, so a larger proportion of bigger fish were collected. When plotted on a frequency histogram the data presented as a negative skew.

The Structure of a Report (page 39)

1. (b) **Methods**: provides the reader with instructions on how the investigation was carried out and what equipment was used. Allows for the procedures to be repeated and confirmed by other investigators.
   (c) **Results:** Provides the findings of the investigation and allows the reader to evaluate these themselves.
   (d) **Discussion:** The findings of the work are discussed in detail so the reader can evaluate the findings. Design limitations, and ways the work could have been improved are also presented.
   (e) **References/Acknowledgments:** Lists sources of information and help used during the investigation. The reader can review the references for more detail if required, and compare your work with other studies in the area of investigation.

2. A poster presents all of the key information from an investigation in an attractive, concise manner which is readily accessible and easy to read. People can quickly determine if the study is of interest to them, and the references provide an opportunity to find out further information if required.

Writing the Methods (page 40)

1. (a)-(d) Any of the following in any order:
   - The transect line should be marked out at intervals, rather than having to 'step out' the distances.
   - Transect length is not given
   - The 100 paces would not be consistent intervals as different people were stepping them out.
   - The biggest plant under the string should not be the one that is recorded; the one directly under the string at a predetermined point (e.g. every 1 m) should be recorded.

Writing Your Results (page 41)

1. Referring to tables and figures in the text clearly indicates which data you are referring to in your synopsis of the results and gives the reader access to these data so that they can assess your interpretation.
2. Tables summarize data and provide a record of the data values, which may not be easily obtained from a graph. Graphs present information in a way that makes any trends or relationships in the data apparent. Such trends may not be evident from the tabulated data. Both formats are valuable for different reasons.

Writing Your Discussion (page 42)

1. Discussion of weaknesses in your study shows that you have considered these and acknowledged them and the effect that they may have had on the outcome of your investigation. It also provides the opportunity for those repeating the investigation (including yourself) to improve on aspects of the design.
2. A critical evaluation shows that you have examined your results carefully in light of the question(s) you asked and your predictions. Objective evaluation enables you to provide reasonable explanations for any unexpected or conflicting results and identify ways in which to improve your study design in future investigations.
3. The conclusion allows you to make a clear statement about your findings, i.e. whether or not the results support your hypothesis. If your results and discussion have been convincing, the reader should be in agreement with the conclusion you make.

Citing and Listing References (page 43)

1. A bibliography lists all sources of information whereas a reference list includes only those sources that are cited in the text. Usually a bibliography is used to compile the final reference list, which appears in the report.
2. Internet articles can be updated as new information becomes available and the original account is revised. It is important that this is noted because people using that source in the future may find information that was unavailable to the author making the original citation.
3. Reference list as follows:
   Ball, P. (1996): Living factories. New Scientist, 2015, 28-31
   Campbell, N. (1993): Biology. Benjamin/Cummings. Ca.
   Cooper, G. (1997): The cell: a molecular approach. ASM Press, Washington DC. pp. 75-85
   Moore, P. (1996): Fuelled for Life. New Scientist, 2012, 1-4
   O'Hare, L. & O'Hare, K. (1996): Food biotechnology. Biological Sciences Review, 8(3), 25.
   Roberts, I. & Taylor, S. (1996): Development of a procedure for purification of a recombinant therapeutic protein. Australasian Biotechnology, 6(2), 93-99.

KEY TERMS: Mix and Match (page 45)

Accuracy (O), Bibliography (Y), Biological drawing (C), Citation (A), Control (G), Controlled variable (K), Data (D), Datalogger (W), Dependent variable (I), Graph (R), Histogram (L), Hypothesis (M), Independent variable (B), Mean (H), Measurement (P), Median (N), Mode (E), Observation (S), Parameter (BB), Precision (Z), Qualitative data (U), Quantitative data (J), Raw data (Q), Report (X), Sample (V), Scientific method (T), Trend (of data) (F), Variable (AA)

The Biochemical Nature of the Cell (page 48)

1. (a) Low viscosity: Water flows through very small spaces and capillaries. It also enables aquatic organisms to move through it without expending a lot of energy.
   (b) Colorless and transparent: Light penetrates tissue and aquatic environments. This property allows photosynthesis to continue at considerable depth.
   (c) Universal solvent: It is the medium for the chemical reactions of life. Water is also the main transport medium in organisms.
   (d) Ice is less dense than water: Ice floats and also insulates the underlying water.
2. (a) Lipids are important as a ready store of concentrated energy (their energy yield per gram is twice that of carbohydrates). They also provide insulation and a medium in which to transport fat-soluble vitamins. Phospholipids are a major component of cellular membranes.
   (b) Carbohydrates are a major component of most plant cells, a ready source of energy, and they are involved in cellular recognition. They can also be changed into fats.
   (c) Proteins are required for growth and repair of cells. They may be structural, catalytic, or have a variety of other functions as well as being able to be converted into fats.
   (d) Nucleic acids, e.g. DNA and RNA, encode the genetic information for the construction and functioning of an organism.

Organic Molecules (page 49)

1. Carbon, hydrogen, and oxygen.
2. Sulfur and nitrogen.
3. Four covalent bonds (valency of 4).
4. A molecular (or chemical) formula shows the numbers and kinds of atoms in a molecule whereas a structural formula is the graphical representation of the molecular structure showing how the atoms are arranged.
5. A functional group is an atom or group of atoms, such as a carboxyl group, that replaces hydrogen in an organic compound and that defines the structure of a

family of compounds and determines the properties of the family.

6. It is an aldehyde.
7. Either one of: amine group (NH_2) or carboxyl group (COOH).
8. The amino acid cysteine has an R group (SH) that can form disulfide bridges with other cysteines to create cross linkages in a polypeptide chain (protein).

Water and Inorganic Ions (page 50)

1.

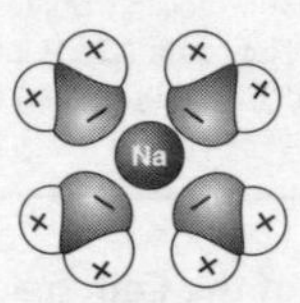

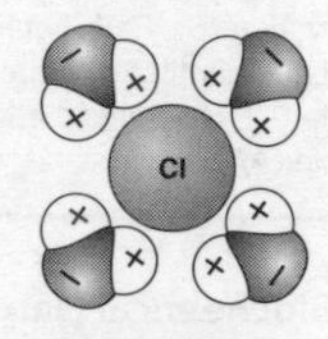

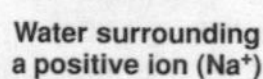

Water surrounding a positive ion (Na^+) Water surrounding a negative ion (Cl^-)

2. The **dipole nature** of water means that it is a good solvent for many substances, e.g. ionic solids and other polar molecules such as sugars and amino acids. It is therefore readily involved in biochemical reactions.
3. Inorganic compounds can be formally defined with reference to what they are not, i.e. organic compounds. Organic compounds are those which contain carbon, with the exception of a few types of carbon containing inorganic compounds such as carbonates, carbon oxides, and cyanides, as well as elemental carbon.
4. (a) Calcium: Calcium ions (Ca^{2+}) are a component of bones and teeth. Ca^{2+} also functions as a biological messenger.
   Deficiency: Depletion of bone stores and increased tendency to bone fracture, disturbance to calcium regulating mechanisms, impairment of nerve and muscle function.
   (b) Iron: Iron ions (Fe^{2+}) are a component of hemoglobin, the main oxygen carrying molecule, where Fe^{2+} is the central ion of the molecule.
   Deficiency: Anemia, fatigue, pallor, irritability, general weakness and breathlessness.
   (c) Phosphorus: Phosphate ions (PO_4^{3-}) are a component of adenosine triphosphate (ATP), an energy-currency molecule which stores energy in an accessible form. Bone is calcium phosphate.
   Deficiency: Anorexia, impaired growth, skeletal demineralization, muscle atrophy and weakness, cardiac arrhythmia, respiratory insufficiency, decreased blood function, nervous system disorders, and even death.
   (d) Sodium: Sodium ions (Na^+) have a role similar to potassium ions in the sodium-potassium pump.
   Deficiency: Electrolyte disturbances and water intoxication (toxic water levels in the blood).
   (e) Sulfur: As part of four amino acids, sulfur is important in a number of the redox reactions of respiration, in carbohydrate metabolism, in protein synthesis, liver function, and in blood clotting. Hydrogen sulfide (H_2S) replaces H_2O in photosynthesis of some bacteria.
   Deficiency: Rare, but sulfur deficiency is attributed to circulatory problems, skin disorders, and various muscle and skeletal dysfunctions.
   (f) Nitrogen: Nitrogen is a constituent element of all living tissues and amino acids.
   Deficiency: Notable in plants with stunting of growth and yellowing of leaves. In animals, nitrogen deficiency manifests as various types of protein deficiency disease, e.g. kwashiorkor, which is characterized by degeneration of the liver, severe anemia, edema, and inflammation of the skin.

Carbohydrates (page 51)

1. **Structural** isomers have the same molecular formula but their atoms are linked in different sequences. For example, fructose and glucose are structural isomers because, although they have the same molecular formula ($C_6H_{12}O_6$), glucose contains an aldehyde group (it is an aldose) and fructose contains a keto group (it is a ketose). In contrast, **optical** isomers are identical in every way except that they are mirror images of each other. The two ring forms of glucose, α and β glucose, are optical isomers, being two mirror image forms.
2. Isomers will have different bonding properties and will form different disaccharides and macromolecules depending on the isomer involved, e.g. glucose and fructose are structural isomers; glucose + glucose forms maltose, glucose + fructose from sucrose. A polysaccharide of the α isomer of glucose forms starch whereas the β isomer forms cellulose.
3. Compound sugars are formed and broken down by condensation and hydrolysis reactions respectively. Condensation reactions join two carbohydrate molecules by a glycosidic bond with the release of a water molecule. Hydrolysis reactions use water to split a carbohydrate molecule into two, where the water molecule is used to provide a hydrogen atom and a hydroxyl group.
4. Cellulose, starch, and glycogen are all polymers of glucose, but differ in form and function because of the optical isomer involved, the length of the polymers, and the degree of branching. **Cellulose** is an unbranched, long chain glucose polymer held by β-1,4 glycosidic bonds. The straight, tightly packed chains give cellulose high tensile strength and resistance to hydrolysis. **Starch** is a mixture of two polysaccharides: amylose (unbranched with α-1,4 glycosidic bonds) and amylopectin (branched with α-1,6 glycosidic bonds). The α-1,4 glycosidic bonds and more branched nature of starch account for its physical properties; starch is powdery and more easily hydrolyzed than cellulose, which exists as tough microfibrils. **Glycogen**, like starch, is a branched polymer. It is similar to amylopectin, being composed of α-glucose molecules, but it is larger and more there are more α-1,6 links. This makes it highly branched, more soluble, and more easily hydrolyzed than starch.

Lipids (page 53)

1. In phospholipids, one of the fatty acids is replaced with a phosphate; the molecule is ionized and the phosphate end is water soluble. Triglycerides are non-polar and not soluble in water.
2. (a) Solid fats: Saturated fatty acids.

(b) Oils: Unsaturated fatty acids.

3. The amphipathic nature of phospholipids (with a polar, hydrophilic end and a hydrophobic, fatty acid end) causes then to orientate in aqueous solutions so that the hydrophobic 'tails' point in together. Hence the bilayer nature of phospholipid membranes.

4. (a) Saturated fatty acids contain the maximum number of hydrogen atoms, whereas unsaturated fatty acids contain some double-bonded carbon atoms.
   (b) Saturated fatty acids tend to produce lipids that are solid at room temperature, whereas lipids that contain a high proportion of unsaturated fatty acids tend to be liquid at room temperature.
   (c) The cellular membranes of an Arctic fish could be expected to contain a higher proportion of unsaturated fatty acids than those of a tropical fish species. This would help them to remain fluid at low temperatures.

5. (a) and (b) any of the following:
   - Male and female sex hormones (testosterone, progesterone, estrogen): regulate reproductive physiology and sexual development.
   - Cortisol: glucocorticoid required for normal carbohydrate metabolism and response to stress.
   - Aldosterone: acts on the kidney to regulate salt (sodium and potassium) balance.
   - Cholesterol is a sterol lipid and, while not a steroid itself, it is a precursor to several steroid hormones and a component of membranes.

6. (a) Energy: Fats provide a compact, easily stored source of energy. Energy yield per gram on oxidation is twice that of carbohydrate.
   (b) Water: Metabolism of lipids releases water (**Note**: Oxidation of triglycerides releases twice as much water as carbohydrate).
   (c) Insulation: Heat does not dissipate easily through fat therefore thick fat insulates against heat loss.

Nucleic Acids (page 55)

1. Labels as follows (only half of the section of DNA illustrated in the workbook is shown):

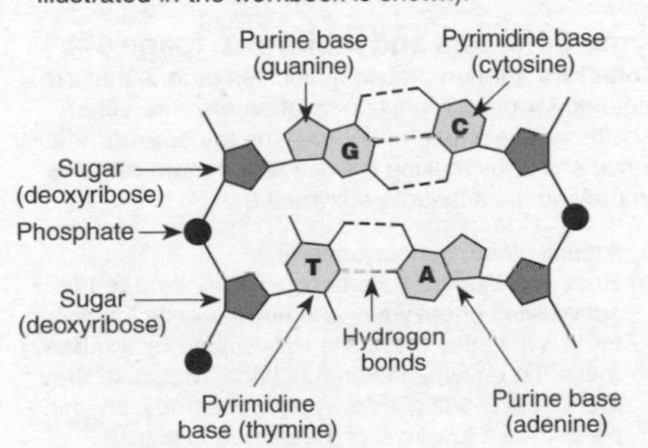

2. (a) The following bases always pair in a normal double strand of DNA: guanine with cytosine, cytosine with guanine, thymine with adenine, adenine with thymine.
   (b) In mRNA, uracil replaces thymine in pairing with adenine.
   (c) The hydrogen bonds in double stranded DNA hold the two DNA strands together.

3. **Nucleotides** are building blocks of nucleic acids (DNA, RNA). Their precise sequence provides the genetic blueprint for the organism.

4. The template strand of DNA is complementary to the coding strand and provides the template for the transcription of the mRNA molecule. The coding strand has the same nucleotide sequence as the mRNA (it carries the code), except that thymine in the coding strand substitutes for uracil in the mRNA.

5.

	DNA	RNA
Sugar present	Deoxyribose	Ribose
Bases present	Adenine Guanine Cytosine Thymine	Adenine Guanine Cytosine Uracil
Number of strands	Two (double)	One (single)
Relative length	Long	Short

Amino Acids (page 57)

1. Twenty different amino acids comprise the building blocks for constructing proteins (which have diverse structural and metabolic functions). The non-protein amino acids have specialised roles as intermediates in metabolic reactions (e.g. as pH buffers) or as the precursors of many important molecules (e.g. neurotransmitters and hormones). Amino acids are also available as dietary supplements for specific purposes.

2. The side chains (R groups) differ in their chemical structure (and therefore their chemical effect).

3. Translation of the genetic code. Genetic instructions from the chromosomes (genes on the DNA) determine the order in which amino acids are joined together.

4. The carboxyl group confers acidic properties to an amino acid, and the R group can also affect the acidic or alkaline nature of the molecule. This means that the amino acid will act as a buffer by removing excess hydroxyl or hydrogen ions present in the surrounding solution. Amino acids retain this capacity even when incorporated into proteins.

5. **Essential amino acids** (isoleucine, leucine, lysine, methionine, phenylalanine, threonine, tryptophan, and valine in adults, and (in children) arginine and histidine as well) cannot be manufactured by the human body, they must be included in the food we eat.

6. **Condensation reactions involve the joining of two amino acids** (or an amino acid to a dipeptide or polypeptide) by a peptide bond with the release of a water molecule. **Hydrolysis involves the splitting of a dipeptide** (or the splitting of an amino acid from a polypeptide) where the peptide bond is broken and a water molecule is used to provide a hydrogen atom and a hydroxyl group.

Proteins (page 59)

1. (a) **Structural**: Proteins form an important component of connective tissues and epidermal structures: collagen, keratin (hair, horn etc.). Proteins are also found scattered on, in, and through cell membranes, but tend to have a regulatory role in this instance. Proteins are also important in maintaining a tightly

coiled structure in a condensed chromosome.
(b) **Regulatory**: **Hormones** such as insulin, adrenaline (modified amino acid), glucagon (peptide) are chemical messengers released from glands to trigger a response in a target tissue. They help maintain homeostasis. **Enzymes** regulate metabolic processes in cells.
(c) **Contractile**: Actin and myosin are structural components of muscle fibers. Using a ratchet system, these two proteins move past each other when energy is supplied.
(d) **Immunological**: Gamma globulins are blood proteins that act as antibodies, targeting antigens (foreign substances and microbes) for immobilization and destruction.
(e) **Transport**: Hemoglobin and myoglobin are proteins that act as carrier molecules for transporting oxygen in the bloodstream of vertebrates. Invertebrates usually have some other type of oxygen carrying molecule in the blood.
(f) **Catalytic**: Enzymes, e.g. amylase, lipase, lactase, trypsin, are involved in the chemical digestion of food. A vast variety of other enzymes are involved in just about every metabolic process in organisms.

2. Denaturation destroys protein function because it involves an irreversible change in the precise tertiary or quaternary structure that confers biological activity. For example, a denatured enzyme protein may not have its reactive sites properly aligned, and will be prevented from attracting the substrate molecule.

3. Any of the following:
   - Globular proteins have a tertiary structure that produces a globular or spherical shape. Fibrous proteins have a tertiary structure that produces long chains or sheets, often with many cross-linkages.
   - The structure of fibrous proteins makes them insoluble in water. The spherical nature of globular proteins makes them water soluble.

4. (a) 21 amino acids (b) 29 amino acids

Enzymes (page 61)

1. Catalysts cause reactions to occur more readily. Enzymes are biological molecules (usually proteins) and allow reactions that would not otherwise take place to proceed, or they speed up a reaction that takes place only slowly. Hence the term, **biological catalyst**. The **active site** is critical to this function, as it is the region where substrate molecules are drawn in and positioned in such a way as to promote the reaction.

2. Catabolism breaks large molecules into smaller ones, e.g digestion, cellular respiration (products=lower PE). Anabolism builds larger molecules from smaller ones, e.g. biosynthetic process (product=higher PE).

3. The lock and key model proposed that the substrate was simply drawn into a closely matching cleft (active site) on the enzyme. In this model, the enzyme's active site was a somewhat passive recipient of the substrate.

4. The induced fit model is a modified version of lock and key, where the substrate fits into the active site, and this initiates a change in the shape of the enzyme's active site so that the reaction can proceed.

5. (a) and (b) in any order, any two of:
   - Deviations from the optimum pH.
   - Excessively high temperature (heating).
   - Treatment with heavy metal ions, urea, organic solvents, or detergents.

   All these agents denature proteins by disrupting the non-covalent bonds maintaining the protein's functional secondary and tertiary structure. The covalent bonds providing the primary structure often remain intact but the protein loses solubility and the functional shape of the protein (its active site) is lost.

6. A mutation could result in a different amino acid being positioned in the polypeptide chain. The final protein may be folded incorrectly (incorrect tertiary and quaternary structure) and lose its biological function. **Note**: If the mutation is silent or in a non-critical region of the enzyme, biological function may not be affected.

Enzyme Reaction Rates (page 63)

1. (a) An increase in enzyme concentration increases reaction rate.
   (b) By manufacturing more or less (increasing or decreasing the rate of protein synthesis).

2. (a) An increase in substrate concentration increases reaction rate to a point. Reaction rate does not continue increasing but levels off as the amount of substrate continues to increase.
   (b) The reaction rate changes because after a certain substrate level the enzymes are fully saturated by substrate and the rate cannot increase any more.

3. (a) An optimum temperature for an enzyme is the temperature at which enzyme activity is maximum.
   (b) Most enzymes perform poorly at low temperatures because chemical reactions occur slowly or not all at low temperatures (enzyme activity will reappear when the temperature increases; usually enzymes are not damaged by moderately low temperatures).

4. (a) Optimum pH: pepsin: 1-2, trypsin: approx. 7.5-8.2, urease: approx. 6.5-7.0.
   (b) The stomach is an acidic environment which is the ideal pH for pepsin.

Enzyme Cofactors and Inhibitors (page 64)

1. **Cofactors** are non-protein molecules or ions that are required for proper functioning of an enzyme either by altering the shape of the enzyme to complete the active site or by making the active site more reactive (improving the substrate-enzyme fit).

2. (a) Arsenic, lead, mercury, cadmium.
   (b) Heavy metals are toxic because they bind to the active sites of enzymes and permanently inactivate them. While the active site is occupied by the heavy metal the enzyme is non-functional. Because they are lost exceedingly slowly from the body, anything other than a low level of these metals is toxic.

3. (a) Examples (any one of): nerve gases, cyanide, DDT, parathion, pyrethrins (insecticides).
   (b) **Nerve gases** deactivate the enzyme acetyl-cholinesterase which is important in the functioning of nerves and muscles (it normally deactivates acetylcholine in synapses and prevents continued over-response of nerve and muscle cells).
   Cyanide poisons the enzyme cytochrome oxidase,

one of the enzymes in the electron transport system. It therefore stops cellular respiration.
DDT and other organochlorines: Inhibitors of key enzymes in the nervous system.
Pyrethrins: Insecticides which inactivate enzymes at the synapses of invertebrates. This has a similar over-excitation effect as nerve gases in mammals.

4. In **competitive** inhibition, the inhibitor competes with the substrate for the enzyme's active site and, once in place, prevents substrate binding. A **noncompetitive** inhibitor does not occupy the active site but binds to some other part of the enzyme, making it less able to perform its function as an effective biological catalyst.

5. Whilst noncompetitive inhibitors reduce the activity of the enzyme and slow down the reaction rate, **allosteric inhibitors** block the active site altogether and prevent its functioning completely.

Applications of Enzymes (page 65)

1. (a) The reaction would not proceed or would require high temperatures or pressures to make it proceed.
   (b) The reaction would proceed only slowly.
   Both consequences add expense to a process.

2. Properties of microbial enzymes that make them suitable industrial catalysts. (a)-(c) any three of:
   - Huge microbial diversity: Important because a wide variety of microbial enzymes are available.
   - Microbial culture is relatively straightforward, so the enzymes are cost effective to produce for industry.
   - The enzymes can be immobilized and so used repeatedly and recovered (thus reducing costs).
   - The enzymes often do not require highly specific conditions in which to operate, so industrial processes can proceed at standard temperatures and pressures. Specific enzymes are also available to catalyze reactions where specific conditions (e.g. high temperatures) are required.

3. Brief answers only (one enzyme example given; there are often others). Students might provide more detail.
   (a) **Chymosin** from GE yeast or bacteria (including *E. coli* (Chy-max in the US) and *Kluyveromyces lactis*).
   (b) Used to coagulate the milk protein, casein.

   (a) **Amyloglucosidases** from GE bacteria.
   (b) Used to speed up the conversion of starch to sugars to get a low-calorie beer. **Note**: **proteases**, from GE microbes are used to modify the proteins from the malt and prevent cloudiness in the finishing stage. These are in addition to the proteases arising naturally from the grain in germination, which solubilize the proteins in the grains and make the amino acids available to the yeast.

   (a) **Pectinases** from the soft-rot bacterium *Erwinnia* or from GE *Aspergillus niger*.
   (b) Breaks down the soluble pectin chains remaining in pressed juice and reduces cloudiness.

   (a) **Citrate synthase** is produced by a mutant strain of the fungus *Aspergillus niger*.
   (b) Citrate synthase (also isocitrate dehydrogenase), catalyses the fermentation of sucrose (under nitrogen limitation) to citric acid, a widely used preservative and pH regulator in the food industry.

   (a) **Proteases** from *Bacillus subtilis*.
   (b) Break the peptide bonds in protein-based stains.

   (a) **Invertase** (sucrase) from *Saccharomyces* spp.
   (b) Converts sucrose to glucose and fructose (invert syrup) to produce a soft center in sweets.

   (a) **Glucose oxidase** from the fungus *Aspergillus niger*.
   (b) Used in medical biosensors for the detection of blood glucose level. Glucose oxidase catalyses the conversion of the glucose to gluconic acid.

   (a) **Proteases** from *Bacillus subtilis*.
   (b) Break the peptide bonds in proteins, and digesting the hair and tissue from animal hides.

   (a) **Lactase** from the bacterium *Kluyveromyces lactis*.
   (b) Converts lactose to glucose and galactose in low lactose dairy products.

   (a) **Ligninases** from white rot fungal species.
   (b) Breaks down the lignin in wood pulp and wood waste. Lignin is a complex molecule and several enzymes, including laccase, lignin peroxidase, and manganese peroxidase are involved.

4. (a) Biosensors use biological material, e.g. an enzyme, to detect the presence or concentration of a particular substance. **Note**: The biological material is immobilized within a semi-conductor. Its activity (in response to the substrate), causes an ion change which is detected by a transducer, amplified, and displayed as a read-out.
   (b) An enzyme that uses alcohol as its substrate (e.g. alcohol dehydrogenase) could be immobilized in the biological recognition layer. The product of its activity (on alcohol) would produce a detectable change, which would be amplified and displayed as a read-out.

5. (a) Microbial proteases are used in the pre-tanning processes of leather manufacture (e.g. removing fat and hair from hides). The ease with which they can be produced and the wide range of enzymes available has reduced the costs of leather treatment since they need not be pure formulations to be effective. Microbial proteases are also safer and more environmentally-friendly than the toxic chemicals traditionally used in tanning (tannery effluent has long been a source of environmental pollution from a range of chemicals including lime, sodium sulfide, salt, and organic solvents and dyes).
   (b) Chymosin from microbial sources has replaced much of the rennet from calves stomachs traditionally used in cheese production. Chymosin produced from GE microbial sources is very pure and therefore more predictable in terms of its activity than its traditional rennet counterpart. Microbial chymosin can also be produced quickly, so fluctuating demands can be easily satisfied. Fewer calves are required when microbial sources of chymosin supplement the traditional sources and the cheese from made using microbial chymosin is acceptable to vegetarians and others who will not eat cheese made with calf rennet. These features have (respectively) improved the efficiency, cost effectiveness, and consumer appeal of the product made with microbial chymosin.
   (c) The use of fungal ligninases to treat wood waste has improved the environmental safety and reduced

pollution over the traditional methods (such as organic sulfur compounds). Fungal treatment of wood waste effectively accelerates natural biodegradation processes and allows these to continue even at cooler temperatures. Wood waste treated in this way is suitable for use as compost or in mushroom production.

Biochemical Tests (page 67)

1. R_f = 15 mm ÷ 33 mm = **0.45**
2. R_f must always be less than one because the substance cannot move further than the solvent front.
3. Chromatography would be an appropriate technique if the sample was very small or when the substance of interest contains a mixture of several different compounds and neither is predominant.
4. Immersion would just wash out the substance into solution instead of separating the components out behind a solvent front.
5. Leucine, arginine, alanine, glycine (most soluble to least soluble).
6. Lipids are insoluble in water. They will not form an emulsion in water unless they have first been dissolved in ethanol (a non-polar solvent).

KEY TERMS: Word Find (page 68)

G	L	Y	C	O	G	E	N	K	Z	I	Z	A	C	T	I	V	E	S	I	T	E	L	I	I
S	U	G	P	G	S	L	I	P	I	D	S	D	S	Z	I	O	P	C	B	Q	M	X	M	S
A	G	F	L	R	T	D	H	E	X	E	R	G	O	N	I	C	H	E	A	Y	E	P	O	O
T	P	H	X	O	I	H	Z	I	Z	A	J	D	C	A	J	Y	O	L	M	V	V	N	L	M
U	E	Y	B	G	B	M	I	I	R	O	N	U	Q	N	O	L	S	L	I	C	X	E	E	E
R	A	N	A	I	Y	U	A	Y	N	I	Z	G	D	A	R	T	P	U	N	H	U	U	C	R
A	F	W	Z	C	O	E	L	R	P	A	I	H	W	B	G	Q	H	L	O	O	Z	T	U	S
T	C	O	X	Y	S	S	Z	A	Y	W	Y	I	G	O	A	Y	O	O	A	Q	I	R	L	D
E	H	S	Y	H	M	L	E	S	R	S	D	W	E	L	N	K	L	S	C	W	N	A	A	M
D	I	T	O	R	N	E	D	N	H	F	T	I	R	I	I	W	I	E	I	A	D	L	R	Q
S	T	A	Z	H	U	X	S	U	S	I	Y	R	P	C	C	B	P	U	D	T	U	F	N	W
G	I	R	G	A	B	O	X	V	S	O	P	R	U	O	Y	J	I	X	S	E	C	A	F	B
O	N	C	M	F	I	B	R	O	U	S	R	S	G	C	L	W	D	L	L	R	E	T	M	A
I	Y	H	C	O	N	D	E	N	S	A	T	I	O	N	T	A	S	O	N	T	D	S	N	H
A	I	C	O	F	A	C	T	O	R	S	I	T	A	X	U	U	R	B	U	M	F	H	S	E
E	F	P	D	I	P	E	P	T	I	D	E	V	R	M	R	Q	R	O	R	O	I	X	E	U
D	E	N	A	T	U	R	A	T	I	O	N	V	J	P	V	C	B	E	X	R	T	Y	J	A

Types of Living Things (page 70)

1. (a) **Autotrophic**: plant cells, some protist(an) cells, some bacterial cells.
   (b) **Heterotrophic**: fungal cells, animal cells, some protist(an) cells, some bacterial cells, viruses.
2. (a) Prokaryotic cells are much smaller (and simpler) than the cells of eukaryotes.
   (b) Prokaryotic cells are bacterial cells while eukaryotic cells are all cell types other than bacteria and viruses. **Note**: More specifically, prokaryotes lack a distinct nucleus, have no membrane-bound organelles, have a cell wall usually containing peptidoglycan and their DNA is present as a single, naked chromosome.
3. (a) Fungi are plant-like in their appearance and habit (growth form, lack of movement, habitat etc.)
   (b) This classification was erroneous because fungi (unlike plants) lack chlorophyll, their cell walls contain chitin (not cellulose), and they are heterotrophic (not autotrophic).
4. Protists often exhibit both animal-like and plant-like features and the group is very diverse in terms of nutrition, reproduction, and structure.
   Note: From a phylogenetic point of view, the protists are not monophyletic and should be classified accordingly.

Bacterial Cells (page 71)

1. (a) Locomotion. Flagella enable bacterial movement.
   (b) Fimbriae are shorter, straighter, and thinner than flagella. They are used for attachment rather than locomotion.
2. (a) Bacterial cell wall lies outside the plasma (cell surface) membrane. It is a semi-rigid structure composed of a macromolecule called peptidoglycan, and contains varying amounts of lipopolysaccharides and lipoproteins.
   (b) The glycocalyx is a viscous, gelatinous layer which lies outside the cell wall. It usually comprises polysaccharide and/or polypeptide, but not peptidoglycan, and may be firmly or loosely attached to the wall.
3. (a) Bacteria usually reproduce by binary fission, where the DNA replicates and the cell then splits into two.
   (b) Conjugation differs from binary fission in that DNA is exchanged between one bacterial cell (the donor) and another (the recipient). The recipient cell gains DNA from the donor.
   (c) Conjugation allows bacteria that have acquired new genes (e.g. a beneficial mutation for antibiotic resistance) to pass on those genes to other (compatible) bacteria. This allows for rapid genetic change since mutations are not lost but are passed on to other bacteria. **Note**: The genes for antibiotic resistance are often carried on extra-chromosomal (plasmid) DNA, so that chances of gene transfer through conjugation are increased.
4. Plasmids are used extensively in recombinant DNA technology. Being accessory to the main chromosome, the plasmid DNA can be manipulated easily. Using restriction enzymes, foreign genes (e.g. gene for producing insulin) can be spliced into a plasmid, which then carries out the instructions of the foreign gene.

Unicellular Eukaryotes (page 73)

1. Summary for each organism under the given headings:
   Amoeba:
   Nutrition: Heterotrophic, food (e.g. bacteria) ingested by phagocytosis. Food digested in food vacuoles.
   Movement: By pseudopodia (cytoplasmic projections).
   Osmoregulation: Contractile vacuole.
   Eyespot: Absent
   Cell wall: Absent

   Paramecium:
   Nutrition: Heterotrophic (feeds on bacteria and small protists). Food digested in food vacuoles.
   Movement: By beating of cilia.
   Osmoregulation: Contractile vacuoles.
   Eyespot: Absent

Cell wall: Absent

Euglena:
Nutrition: Autotrophic (photosynthetic), but can be heterotrophic when light deprived.
Movement: By flagella (one larger, which is labeled, and one very small, beside the gullet).
Osmoregulation: Contractile vacuole.
Eyespot: Present.
Cell wall: Absent, although there is a wall-like pellicle, which lies inside the plasma membrane and is flexible.

Chlamydomonas:
Nutrition: Autotrophic (photosynthetic).
Movement: By flagella.
Osmoregulation: Contractile vacuole.
Eyespot: Present.
Cell wall: Present.

2. *Amoeba, Paramecium, Euglena, Chlamydomonas.*
3. An eye spot enables an autotroph to detect light so that it can move into (well lit) regions where it can photosynthesize.

Fungal Cells (page 74)

1. *In any order:* Cell wall of chitin; consist of filaments of cells (hyphae) that may form a network or mycelium; all are chemoheterotrophs, reproduction by spores.
2. Diploid yeast cells reproduce asexually by fission or budding (not involving spore formation). Sexual reproduction in yeasts involves the production of haploid ascospores which fuse to produce new diploid yeast cells. Filamentous fungi (molds such as *Rhizopus*) have a sexual phase characterized by fusion of gametes to produce a **zygospore** and an asexual phase characterized by production of asexual, haploid spores from a sporangium.
3. Many examples are possible: (a) and (b) any of:
   - *Pencillium* mold: production of antibiotics.
   - *Streptomyces* mold: production of antibiotics.
   - *Saccharomyces* spp. esp. *Saccharomyces cerevisiae*: production of alcoholic beverages.
   - *Aspergillus* spp.: production of industrial enzymes.

Plant Cells (page 75)

1. A: Nucleus C: Nucleus
   B: Cell wall D: Chloroplasts
2. (a) Cytoplasmic streaming is the rapid movement of cytoplasm within eukaryotic cells, seen most clearly in plant and algal cells.

   (b)

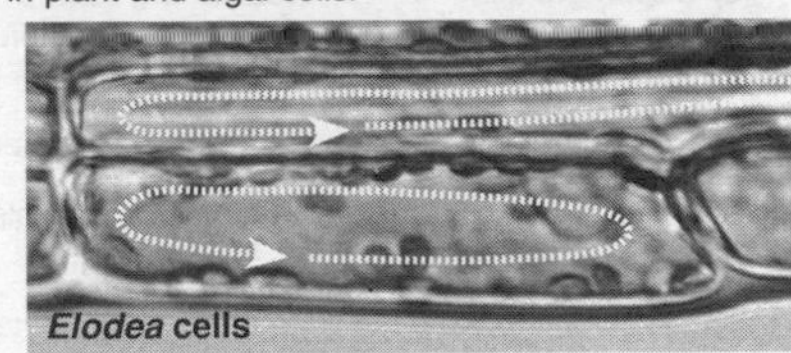

3. (a)-(c) any three of:
   - Starch (branched carbohydrate) granules stored in amyloplasts (energy store)
   - Chloroplasts, discrete plastids containing the pigment chlorophyll, involved in photosynthesis.
   - Large vacuole, often central (vacuoles are present in animal cells, but are only small).
   - Cell wall of cellulose forming the rigid, supporting structure outside the plasma membrane.
   - Plasmodesmata

Animal Cells (page 76)

1. A: Nucleus
   B: Plasma membrane
   C: Nucleus
2. (a)

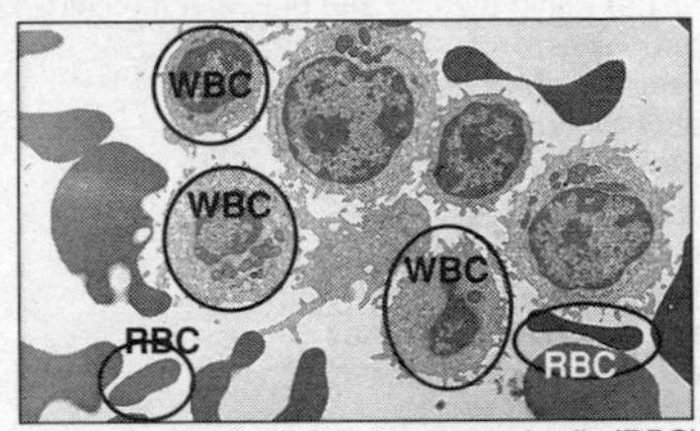

White blood cells (WBC) & red blood cells (RBC)

   (b) Any of the following reasons: The RBCs have no nucleus and they are smaller than the white blood cells. The white blood cells have extensions of the plasma membrane (associated with being mobile and phagocytic), are larger than the RBCs, and have a nucleus.
3. Centrioles (although these are present in lower plants, they are absent from higher plants). They are microtubular structures responsible for forming the poles and the spindles during cell division.

Cell Sizes (page 77)

1.

(a) *Amoeba*:	300 μm	0.3 mm
(b) Foraminiferan:	400 μm	0.4 mm
(c) *Leptospira*:	7-8 μm	0.007-0.008 mm
(d) Epidermis:	120 μm	0.12 mm
(e) *Daphnia*:	2500 μm	2.5 mm
(f) *Papillomavirus*:	0.13 μm	0.00013 mm

2. *Papillomavirus, Leptospira*, Epidermis, *Amoeba*, Foraminiferan, *Daphnia*
3. Epidermis (possibly), *Amoeba*, Foraminiferan, *Daphnia*
4. (a) 0.00025 mm (b) 0.45 mm (c) 0.0002 mm

Cell Structures and Organelles (page 78)

(b) **Name**: Ribosome
Location: Free in cytoplasm or bound to rough ER
Function: Synthesize polypeptides (=proteins)
Present in plant cells: Yes
Present in animal cells: Yes
Visible under LM: No

(c) **Name**: Mitochondrion
Location: In cytoplasm as discrete organelles
Function: Site of cellular respiration (ATP formation)
Present in plant cells: Yes
Present in animal cells: Yes
Visible under LM: Not with most standard school LM, but can be seen using high quality, high power LM.

(d) **Name**: Golgi apparatus

Location: In cytoplasm associated with the smooth endoplasmic reticulum, often close to the nucleus.
Function: Final modification of proteins and lipids. Sorting and storage for use in the cell or packaging molecules for export.
Present in plant cells: Yes
Present in animal cells: Yes
Visible under LM: Not with most standard school LM, but may be visible using high quality, high power LM.

(e) **Name**: Endoplasmic reticulum (in this case, rough ER)
Location: Penetrates the whole cytoplasm
Function: Involved in the transport of materials (e.g. proteins) within the cell and between the cell and its surroundings.
Present in plant cells: Yes
Present in animal cells: Yes
Visible under LM: No

(f) **Name**: Chloroplast
Location: Within the cytoplasm
Function: The site of photosynthesis
Present in plant cells: Yes
Present in animal cells: No
Visible under LM: Yes

(g) **Name**: Cytoskeleton
Location: Throughout cytoplasm
Function: Provides structure and shape to a cell, responsible for cell movement (e.g. during muscle contraction), and provides intracellular transport of organelles and other structures.
Present in plant cells: Yes
Present in animal cells: Yes
Visible under LM: No

(h) **Name**: Cellulose cell wall
Location: Surrounds the cell and lies outside the plasma membrane.
Function: Provides rigidity and strength, and supports the cell against changes in turgor.
Present in plant cells: Yes
Present in animal cells: No
Visible under LM: Yes

(i) **Name**: Cell junctions (an animal example is given)
Location: At cell membrane surface, connecting adjacent cells.
Function: Depends on junction type. Desmosomes fasten cells together, gap junctions act as communication channels between cells, and tight junctions prevent leakage of extracellular fluid from layers of epithelial cells.
Present in plant cells: Yes, as plasmodesmata
Present in animal cells: Yes
Visible under LM: No

(j) **Name**: Lysosome and food vaculoe (given)
Lysosome
Location: Free in cytoplasm.
Function: Ingests and destroys foreign material. Able to digest the cell itself under some circumstances.
Present in plant cells: Yes but variably (vacuoles may have a lysosomal function in some plant cells).
Present in animal cells: Yes
Visible under LM: No

Vacuole (a food vacuole in an animal cell is shown, so students may answer with respect to this).
Location: In cytoplasm.
Function: In plant cells, the vacuole (often only one) is a large fluid filled structure involved in storage and support (turgor). In animal cells, vacuoles are smaller and more numerous, and are involved in storage (of water, wastes, and soluble pigments).
Present in plant cells: Yes, as (a) large structure(s).
Present in animal cells: Yes, smaller, more numerous
Visible under LM: Yes in plant cells, no in animal cells.

(k) **Name**: Nucleus
Location: Discrete organelle, position is variable.
Function: The control center of the cell; the site of the nuclear material (DNA).
Present in plant cells: Yes
Present in animal cells: Yes
Visible under LM: Yes.

(l) **Name**: Centrioles
Location: In cytoplasm, usually next to the nucleus.
Function: Involved in cell division (probably in the organization of the spindle fibers).
Present in plant cells: Variably (absent in higher plants)
Present in animal cells: Yes
Visible under LM: No.

(m) **Name**: Cilia and flagella (given)
Location: Anchored in the cell membrane and extending outside the cell.
Function: Motility.
Present in plant cells: No
Present in animal cells: Yes
Visible under LM: Variably (depends on magnification and preparation/fixation of material).

Differential Centrifugation (page 81)

1. Cell organelles have different densities and spin down at different rates. Smaller organelles take longer to spin down and require a higher centrifugation speed to separate out.
2. The sample is homogenized (broken up) before centrifugation to rupture the cell surface membrane, break open the cell, and release the cell contents.
3. (a) Isotonic solution is needed so that there are no volume changes in the organelles.
   (b) Cool solution prevent self digestion of the organelles by enzymes released during homogenization.
   (c) Buffered solution prevents pH changes that might denature enzymes and other proteins.
4. (a) Ribosomes and endoplasmic reticulum
   (b) Lysosomes and mitochondria
   (c) Nuclei

Identifying Cell Structures (page 82)

1. (a) Cytoplasm
   (b) Vacuole
   (c) Starch granule
   (d) Chloroplast
   (e) Mitochondria
   (f) Cell wall
   (g) Chromosome
   (h) Nuclear membrane
   (i) Endoplasmic reticulum
   (j) Plasma membrane
2. 9 cells (1 complete cell, plus the edges of 8 others)
3. Plant cell; it has chloroplasts and a cell wall. It also has a highly geometric cell shape.
4. (a) Cytoplasm located between the plasma membrane and nuclear membrane (extranuclear).
   (b) Composition of cytoplasm: A watery soup of dissolved substances. In eukaryotic cells, organelles are found in the cytoplasm. Cytoplasm = cytosol (including cytoskeleton) + organelles.
   (c) Prokaryotic cells do not have a defined nucleus.

5. (a) Starch granules, which occur within specialized plastids called leukoplasts. Starch granules are non-living inclusions, deposited as a reserve energy store.
   (b) Vacuoles, which are fluid filled cavities bounded by a single membrane. Plant vacuoles contain cell sap; an aqueous solution of dissolved food material, ions, waste products, and pigments. **Note**: Young plant cells (such as the one pictured) usually have several small vacuoles, which unite in a mature cell to form a large, permanent central vacuole.

Optical Microscopes (page 83)

1. Compound microscope (a)-(h) and dissecting microscope (i)-(m) as follows:
   (a) Eyepiece lens
   (b) Arm
   (c) Coarse focus knob
   (d) Fine focus knob
   (e) Objective lens
   (f) Mechanical stage
   (g) Condenser
   (h) In-built light source
   (i) Eyepiece lens
   (j) Eyepiece focus
   (k) Focus knob
   (l) Objective lens
   (m) Stage

2. Phase contrast: used where the specimen is transparent (to increase contrast between transparent structures). **Note**: It is superior to dark field because a better image of the interior of specimens is obtained.

3. (a) Plant cell, any two of: cell wall, nucleus (may see chromatin if stained appropriately), vacuole, cell membrane (high magnification), Golgi apparatus, mitochondria (high magnification), chloroplast, cytoplasm (if stained), nuclear envelope (maybe).
   (b) Animal cell, any two of: nucleus (may see chromatin if stained appropriately), centriole, cell membrane (high magnification), Golgi apparatus, mitochondria (high magnification), cytoplasm (if stained), nuclear envelope (maybe).

4. Any one of: Ribosomes, microtubules, endoplasmic reticulum, Golgi vesicles (free), nuclear envelope as two layers, lysosomes (animal cells). Also detail of organelles such as mitochondria and chloroplasts.

5. (a) Leishman's stain
   (b) Schultz's solution / iodine solution
   (c) Schultz's solution
   (d) Aniline sulfate / Schultz's solution
   (e) Methylene blue
   (f) Schultz's solution

6. (a) 600X magnification (b) 600X magnification

7. Bright field microscopes produces a flat (2-dimensional) image, which looks through a thin, transparent sample. Dissecting microscopes produces a 3-dimensional image, which looks at the surface details.

8. **Magnification** is the number of times larger an image is than the specimen. **Resolution** is the degree of detail which can be achieved. The limit of resolution is the minimum distance by which two points in a specimen can be separated and still be distinguished as separate points. **Note**: By adding stronger, or more, lenses, a light microscope can magnify an image many thousands of times but its resolution is limited. EMs have a greater resolving power than LMs because of the very short wavelength of the electrons used.

Electron Microscopes (page 85)

1. The limit of resolution (see #8 previously) is related to wavelength (about 0.45X the wavelength). The shortest visible light has a wavelength of about 450 nm giving a resolution of 0.45 x 450 nm; close to 200 nm. Points less than 200 nm apart will be perceived as one point or a blur. Electron beams have a much shorter wavelength than light so the resolution is much greater (points that are 0.5 nm apart can be distinguished as separate points; a resolving power 400X that of a LM).

2. (a) TEM: Used to (any of): show cell ultrastructure i.e organelles; to investigate changes in the number, size, shape, or condition of cells and organelles i.e. demonstrate cellular processes or activities; to detect the presence of viruses in cells.
   (b) SEM: Used to (any of): show the surface features of cells, e.g. guard cell surrounding a stoma; to show the surface features of entire organisms for identification purposes (often used for invertebrates and viruses); for general identification by surface feature, e.g. pollen grains in paleoclimate research.
   (c) Bright field (compound): Used for (any of): examining prepared sections of tissue for cellular detail; for examining living tissue for large scale movements, e.g. blood flow in capillaries or cytoplasmic streaming.
   (d) Dissecting: Used for (any of): examining living specimens for surface detail and structures; sorting material from samples (e.g. leaf litter or stream invertebrates; dissecting a small organism where greater resolution than the naked eye is required.

3. A TEM
   B Bright field LM
   C TEM
   D Bright field LM
   E SEM
   F Bright field LM
   G Dissecting LM
   H SEM

Interpreting Electron Micrographs (page 87)

1. (a) Chloroplast
   (b) Plant cells, particularly in leaf and green stems.
   (c) Function: Site of photosynthesis. Captures solar energy to build glucose from carbon dioxide and water.
   (d) Any two of the following:

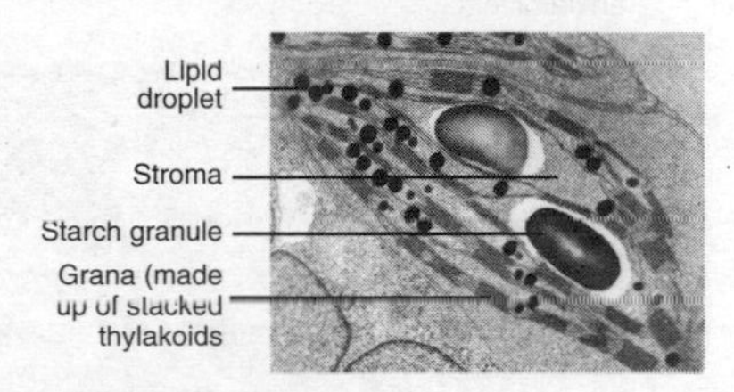

2. (a) Golgi apparatus
   (b) Plant and animal cells
   (c) Function: Packages substances to be secreted by the cell. Forms a membrane vesicle containing the chemicals for export from the cell (e.g. nerve cells export neurotransmitters; endocrine glands export hormones; digestive gland cells export enzymes).

3. (a) Mitochondrion
   (b) Plant and animal cells (most common in cells that have high energy demands, such as muscle).

(c) Function: Site of most cellular respiration, which releases energy from food (glucose) to fuel most cellular reactions (i.e. metabolism).

(d)

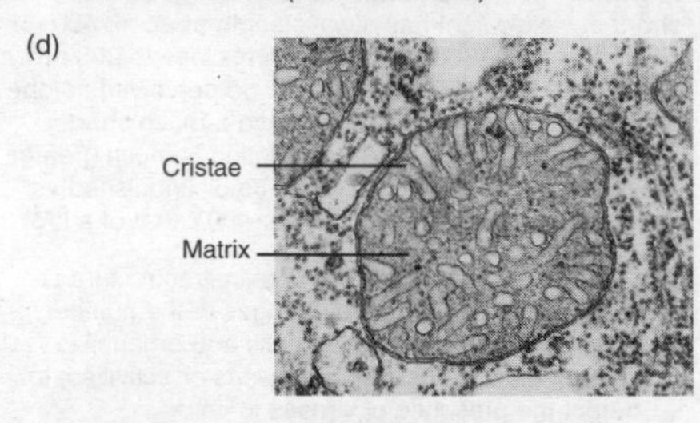

4. (a) Endoplasmic reticulum
   (b) Plant and animal cells (eukaryotes)
   (c) Function: Site of protein synthesis (translation stage). Transport network that moves substances through its system of tubes. Many complex reactions need to take place on the surface of the membranes.
   (d) Endoplasmic reticulum structure: ribosomes.

5. (a) Nucleus
   (b) Plant and animal cells (eukaryotes)
   (c) Function: Controls cell metabolism (all the life-giving chemical reactions), and functioning of the whole organism. These instructions are inherited from one generation to the next.

(d)

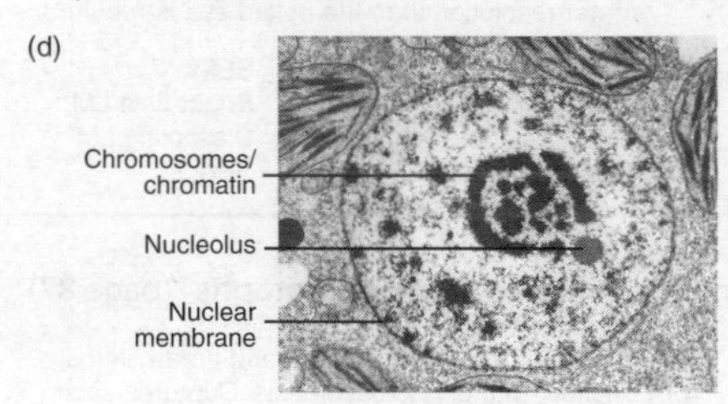

6. (a) Function: Controls the entry and exit of substances into and out of the cell. Maintains a constant internal environment.

(b)

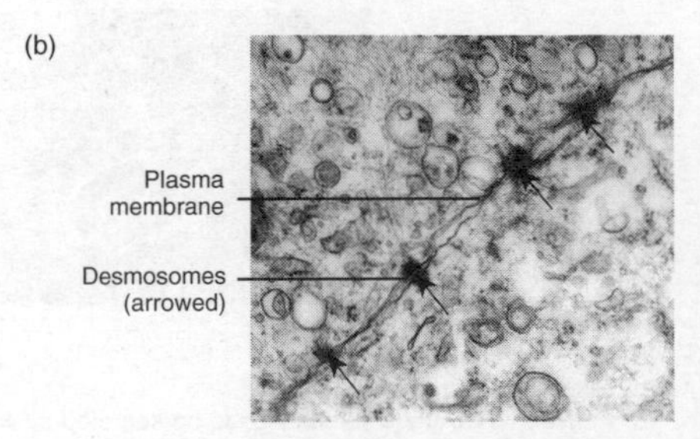

7. Generalized cell:

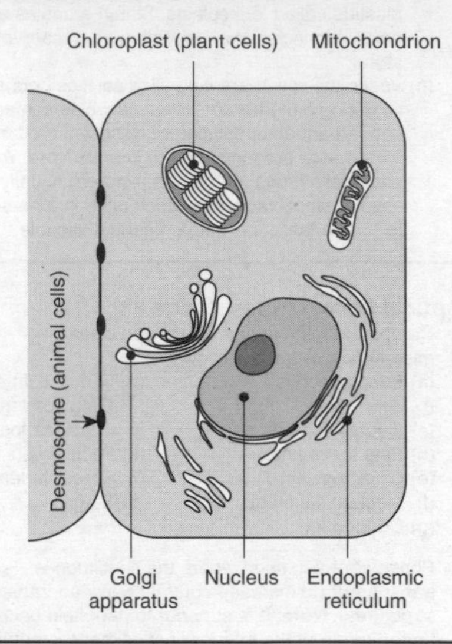

Cell Processes (page 92)

1. (a) Golgi apparatus
   (b) Cytoplasm, mitochondria
   (c) Plasma membrane, vacuoles
   (d) Plasma membrane, vacuoles
   (e) Endoplasmic reticulum, ribosomes, nucleus
   (f) Chloroplasts
   (g) Centrioles, nucleus
   (h) Lysosomes
   (i) Plasma membrane, Golgi apparatus

2. **Metabolism** describes all the chemical processes of life taking place inside the cell. Examples include cellular respiration, fatty acid oxidation, photosynthesis, digestion, urea cycle, and protein synthesis.

The Structure of Membranes (page 93)

1. (a) Channel and carrier proteins
   (b) Carrier proteins
   (c) Glycoproteins, glycolipids
   (d) Cholesterol

2. (a) Membranes are composed of a phospholipid bilayer in which are embedded proteins, glycoproteins, and glycolipids. The structure is relatively fluid and the proteins are able to move within this fluid matrix.
   (b) This model accounts for the properties we observe in cellular membranes: its fluidity (how its shape is not static and how its components move within the membrane, relative to one-another) and its mosaic nature (the way in which the relative proportions of the membrane components, i.e. proteins, glycoproteins, glycolipids etc, can vary from membrane to membrane). The fluid mosaic model also accounts for how membranes can allow for the selective passage of materials (through protein channels for example) and how they enable cell-cell recognition (again, as a result of membrane components such as glycoproteins).

3. The plasma membrane forms the outer limit of the cell, contains proteins that confer cellular recognition and controls the entry and exit of materials in and out of the cell. Intracellular membranes keep the cytoplasm separate from the extracellular spaces and provide compartments for localization of metabolic reactions and a surface for attaching enzymes.

4. (a) Golgi, mitochondria, chloroplasts, nucleus.
   (b) In general, membranes compartmentalize the location of reactions, control the entry and exit of substances and/or provides an enzyme attachment surface.

5. (a) Cholesterol lies between the phospholipids and prevents close packing. It functions to keep membranes more fluid. The greater the amount of cholesterol in the membrane the greater its fluidity.
   (b) At temperatures close to freezing, high proportions of membrane cholesterol is important in keeping membranes fluid and functioning.

6. (a)-(c) in any order: oxygen, food (glucose), minerals and trace elements, water.

7. (a) Carbon dioxide (b) Nitrogenous wastes

8. Plasma membrane:

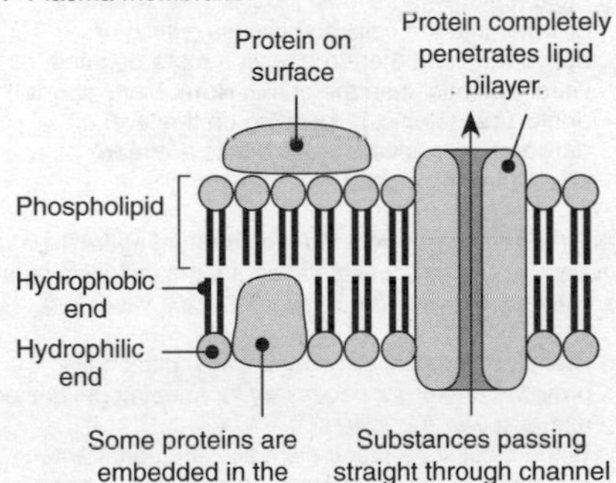

The Role of Membranes in Cells (page 95)

1. Membrane systems and organelles provide compartments within cells which allow enzymatic reactions in the cell to be localized. This achieves greater efficiency in cell function and keeps potentially harmful reactions and substances (e.g. hydrogen peroxide) contained.

2. (a) Golgi apparatus
   (b) lysosome
   (c) mitochondrion
   (d) rough endoplasmic reticulum
   (e) smooth endoplasmic reticulum
   (f) chloroplast

3. Membrane surface area is increased within cells and organelles by invaginations or by a long flattened shape which increases the surface area to volume ratio.

4. (a) High membrane surface area provides a greater area over which membrane-bound reactions can occur. This increases the speed and efficiency with which metabolic reactions can take place.
   (b) Channel and carrier proteins facilitate selective transport of substances through membranes. They can help to speed up the transport of substances into and out of the cell, especially for enzymatic reactions requiring a steady supply of substrate and constant removal of end-product, e.g. ADP supply to the mitochondrion during cellular respiration.

5. (a) Non-polar (lipid-soluble) molecules are able to dissolve in the lipid bilayer structure of the membrane and diffuse easily into the cell whereas the polar (lipid-insoluble) molecules have to be actively transported through the membrane.
   (b) Transportation of lipid-soluble molecules by diffusion alone into a cell is much quicker than that of lipid-insoluble molecules that have to be actively transported across the plasma membrane. This also increases the speed and efficiency with which metabolic reactions can take place.

How Do We Know? Membrane Structure (page 97)

1. Fixing a sample and then freezing it preserves its structure and allows it to be cleaved. The cleavage (splitting) process reveals internal structure. The techniques used in freeze fracture (splitting a lipid bilayer) and electron microscopy (coating in metals) reveal how proteins are organised in the membrane.

2. The impressions left in one side of the membrane after freeze fracture give evidence of proteins located within the membrane. Fracturing the membrane allowed scientists to observe the presence of integral membrane proteins which span the membrane lipid bilayer. This supported the fluid mosaic model, in which membrane bound proteins are able to move relatively freely within the membrane.

3. If the bilayer had a continuous protein coat, the freeze fracture specimen would look flat and uniform when viewed under the electron microscope.
   The proteins are discrete complexes randomly spaced throughout the lipid bilayer. The bumps observed are where the proteins were located.

Modification of Proteins (page 98)

1. (a) **Glycoproteins** are proteins with attached carbohydrates (often relatively small polymers of sugar units).
   (b) In any order, three roles of glycoproteins:
   - **Intercellular recognition**: Present on cell surfaces for recognition between cells (when cells interact to form tissues and for immune function).
   - **Transport**: Embedded in cell membranes to transport molecules through the membrane (the sugars help to maintain the position of the glycoprotein in the membrane).
   - **Regulation**: Secretory proteins from glands with a role in regulation, e.g. many pituitary hormones.

2. (a) **Lipoproteins** are proteins with attached fatty acid molecules.
   (b) Lipoproteins transport lipid molecules in the plasma between different organs in the body.

3. Proteins made on free ribosomes are released directly

into the cytoplasm; there is no facility for attachment of carbohydrate as this generally requires a packaging region (the Golgi apparatus).

4. Protein orientation in the membrane is important because it is usually critical to the functional role of the protein, e.g. in intercellular recognition or transport.

Packaging Macromolecules (page 99)

1. Large organic polymers made up of many repeating units of smaller molecules. They have a high molecular weight. Examples include proteins and nucleic acids.

2. Polypeptides are synthesized by membrane bound ribosomes so that they can be easily threaded through the ER membrane into the cisternal space of the ER. Here they are in place for subsequent modification, packaging and export.

3. The carbohydrates attached to glycoproteins aid in the recognition or function of the protein so that it is transported to the correct destination and performs its appropriate functional role.

4. (a) Rough ER: Ribosomes on the rough ER assemble the proteins destined for secretion.
   (b) Smooth ER: Synthesis of lipids, e.g. steroid hormones and phospholipids, and packages them into transport vesicles.
   (c) Golgi apparatus: Receives transport vesicles. Modifies, stores, and transports molecules for export around or from the cell.
   (d) Transport vesicles: These bud off the ER and move substances to the Golgi apparatus.

Diffusion (page 100)

1. (a) Large surface area (b) Thin membrane

2. Concentration gradients are maintained by (any one of):
   - Constant use or transport away of a substance on one side of a membrane (e.g. use of ADP in mitochondria).
   - Production of a substance on one side of a membrane (e.g. production of CO_2 by respiring cells).

3. Ionophores allow the preferential passage of some molecules but not others.

Osmosis and Water Potential (page 101)

1. Zero

2. and 3

(a)

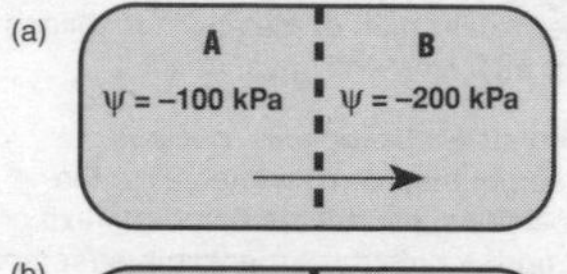

(b)

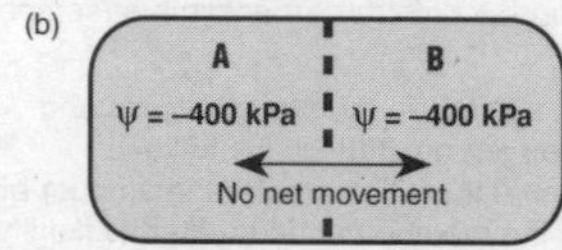

(c)

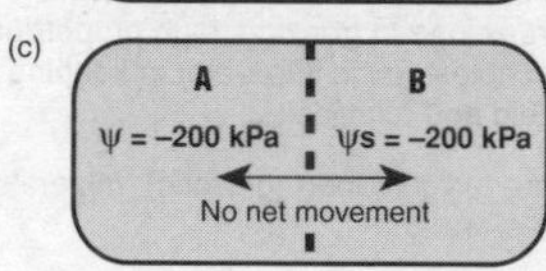

4. (a) Hypotonic
   (b) Fluid replacements must induce the movement of water into the cells and tissues (which are dehydrated and therefore have a more negative water potential than the drink). **Note**: Many sports drinks are isotonic. Depending on the level of dehydration involved, these drinks are more effective when diluted.

5. *Paramecium* is hypertonic to the surrounding freshwater environment; water constantly enters the cell. This must be continually pumped out (by contractile vacuoles).

6. (a) Pressure potential generated within plant cells provides the turgor necessary for keeping unlignified plant tissues supported.
   (b) Without cell turgor, soft plant tissues (soft stems and flower parts for example) would lose support and wilt. **Note** that some tissues are supported by structural components such as lignin.

7. Animal cells are less robust than plant cells against changes in net water content: Excess influx will cause bursting and excess loss causes crenulation.

8. (a) Water will move into the cell and it will burst (lyse).
   (b) The cell would lose water and the plasma membrane would crinkle up (crenulate).
   (c) Water will move into the cell and it will burst (lyse).

9. Malarial parasite: **isotonic** to blood.

Surface Area and Volume (page 103)

1.

Cube	Surface area	Volume	Ratio
3 cm:	3 x 3 x 6 = 54	3 x 3 x 3 = 27	2.0 to 1
4 cm:	4 x 4 x 6 = 96	4 x 4 x 4 = 64	1.5 to 1
5 cm:	5 x 5 x 6 = 150	5 x 5 x 5 = 125	1.2 to 1

2. Surface area to volume graph:

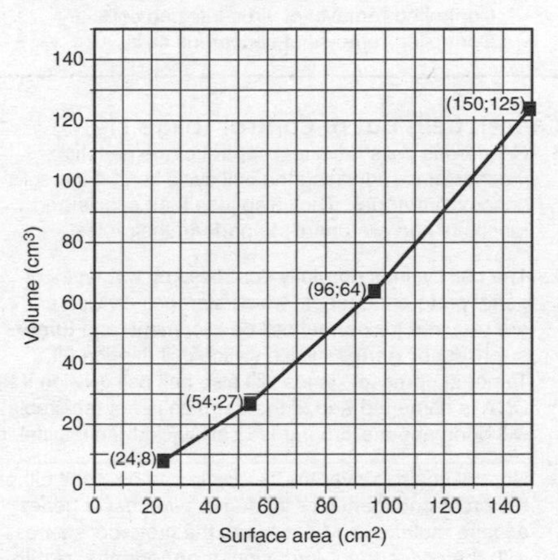

3. Volume
4. Increasing size leads to less surface area for a given volume. The surface area to volume ratio decreases.
5. Less surface area at the cell surface. This is the gas exchange surface, therefore large cells will have difficulty moving materials in and out of the cell in the amounts required to meet demands. This is what limits a cell's maximum size. **Note**: Eukaryote cells are typically about 0.01-0.1 mm in size, but some can be bigger than 1 mm. The largest cell is the female sex cell (ovum) of the ostrich, which averages 15-20 cm in length. Technically a single cell, it is atypical in size because almost the entire mass of the egg is food reserve in the form of yolk, which is not part of the functioning structure of the cell itself.

Ion Pumps (page 105)

1. ATP (directly or indirectly) supplies the energy to move substances against their concentration gradient.
2. (a) Cotransport describes coupling the movement of a molecule (such as sucrose or glucose) against its concentration gradient to the diffusion of an ion (e.g. H^+ or Na^+) down its concentration gradient. Note: An energy requiring ion exchange pump is used to establish this concentration gradient.
   (b) In the gut, a gradient in sodium ions is used to drive the transport of glucose across the epithelium. A Na^+/K^+ pump (requiring ATP) establishes an unequal concentration of Na^+ across the membrane. A specific membrane protein then couples the return of Na^+ down its concentration gradient to the transport of glucose (at a rate that is higher than could occur by diffusion alone).
   (c) The glucose diffuses from the epithelial cells of the gut into the blood, where it is transported away. This maintains a low level in the intestinal epithelial cells.
3. Extracellular accumulation of Na^+ (any two of):
   - maintains the gradient that is used to cotransport useful molecules, such as glucose, into cells.
   - maintains cell volume by creating an osmotic gradient that drives the absorption of water
   - establishes and maintains resting potential in nerve and muscle cells
   - provides the driving force for several facilitated membrane transport proteins

Exocytosis and Endocytosis (page 106)

1. **Phagocytosis** is the engulfment of solid material by endocytosis whereas **pinocytosis** is the uptake of liquids or fine suspensions by endocytosis.
2. Phagocytosis examples (any of):
   - Feeding in *Amoeba* by engulfment of material using cytoplasmic extensions called pseudopodia.
   - Ingestion of old red blood cells by Kupffer cells in the liver.
   - Ingestion of bacteria and cell debris by neutrophils and macrophages (phagocytic white blood cells).
3. Exocytosis examples (any of):
   - Secretion from specialized cells in multicellular organisms, e.g. hormones from endocrine cells, digestive secretions from exocrine cells.
   - Expulsion of wastes from unicellular organisms e.g. *Paramecium* and *Amoeba* expelling residues from food vacuoles.
4. Any type of cytosis (unlike diffusion) is an active process involving the use of ATP. Low oxygen inhibits oxidative metabolism and lowers the energy yield from the respiration of substrates (ATP availability drops).
5. (a) **Oxygen**: diffusion.
   (b) **Cellular debris**: phagocytosis.
   (c) **Water**: osmosis.
   (d) **Glucose**: facilitated diffusion.

Active and Passive Transport (page 107)

1. **Passive transport** requires no energy input from the cell; materials follow a concentration gradient.
   Active transport requires considerable amounts of energy (ATP) to make materials go in a direction they would not normally go (at least at the rate required).
2. Gases moving by diffusion: oxygen, carbon dioxide.
3. (Any one of): Cells in the digestive (exocrine) glands of the stomach, pancreas, upper small intestine (duodenum); endocrine glands (e.g. adrenal glands); salivary glands.
4. (a) Protozoan (any one of): *Amoeba, Paramecium*
   (b) In *Paramecium*, a food vacuole develops at the end of the oral groove and is pinched off to circulate within the cell. In *Amoeba*, the pseudopodia engulf a food particle and a vacuole is formed where the membrane pinches off after the particle is engulfed.
   (c) Human cell: Phagocyte (white blood cell).
   (d) The cell has a role in internal defense, ingesting and destroying foreign particles and pathogens such as bacteria.

Cell Division (page 108)

1. (a) Mitosis: Cell division for growth and repair produces cells with 2N chromosome number.
   (b) Meiosis: Cell division for producing gametes (sperm, eggs) with 1N chromosome number.
2. In **spermatogenesis**, the nucleus of the germ cell divides twice to produce four similar sized gametes (sperm cells). In **oogenesis**, the two divisions are not equal and only one of the four nuclei (and most of the cytoplasm) produce the egg cell.
3. In humans, gametes are produced by meiotic division from diploid cells. In plants, gametes are produced by mitosis from haploid gametophytes.

Mitosis and the Cell Cycle (page 109)

1. A. Anaphase
   B. Prophase
   C. Late metaphase (early anaphase is also acceptable).
   D. Late anaphase
   E. Cytokinesis (late telophase is also acceptable).
2. Replicate the DNA to form a second chromatid. Coil up into visible chromosomes to avoid tangling.
3. A. Interphase: The stage between cell divisions (mitoses). Just before mitosis, the DNA is replicated to form an extra copy of each chromosome (still part of the same chromosome as an extra chromatid).
   B. Late prophase: Chromosomes condense (coil and fold up) into visible form. Centrioles move to opposite ends of the cell.
   C. Metaphase: Spindle fibers form between the centrioles. Chromosomes attach to the spindle fibers at the cell 'equator'.
   D. Late anaphase: Chromatids from each chromosome are pulled apart and move in opposite directions, towards the centrioles.
   E. Telophase: Chromosomes begin to unwind again. Two new nuclei form. The cell plate forms across the midline where the new cell wall will form.
   F. Cytokinesis: Cell cytoplasm divides to create two distinct 'daughter cells' from the original cell. It is in this form for most of its existence, and carries out its designated role (normal function).

Apoptosis: Programmed Cell Death (page 111)

1. Syndactyly arises when the rate of apoptosis in the developing limb has been too low to remove the tissue between the digits before the close of the developmental sequence (hence the lack of differentiation between the two toes, which remain partly joined).
2. Any one of the reasons from the following explanation: Apoptosis is a carefully regulated process, which occurs in response to particular signalling factors. Necrosis, in contrast, is the result of traumatic damage. Apoptosis results in membrane-bound cell fragments whereas necrosis results in lysing and spillage of cell contents. Apoptosis results in cell shrinkage and contraction of the chromatin, which does not occur in necrosis.
3. Roles of apoptosis, (a) and (b) any two of the following in any order:
   - Resorption of the larval tail during amphibian metamorphosis.
   - Sloughing of the endometrium during menstruation.
   - The formation of the proper connections (synapses) between neurons in the brain (this requires that surplus cells be eliminated by apoptosis).
   - Controlled removal of virus-infected cells.
   - Controlled removal of cancerous cells.

Cancer: Cells out of Control (page 112)

1. Cancerous cells have lost control of the genetic mechanisms regulating the cell cycle so that the cells become 'immortal'. They also lose their specialized functions and are unable to perform their roles.
2. The cell cycle is normally controlled by two types of gene: proto-oncogenes, which start cell division and are required for normal cell development, and **tumor-suppressor genes**, which switch cell division off. Tumor suppressor genes will also halt cell division if the DNA is damaged and, if the damage is not repairable, will bring about a programed cell suicide (apoptosis).
3. Normal controls over the cell cycle can be lost if either the proto-oncogenes or the tumor suppressor genes acquire mutations. Mutations to the proto-oncogenes, with the consequent formation of **oncogenes**, results in uncontrolled cell division. Mutations to the tumor-suppressor genes results in a failure to regulate the cell repair processes and a failure of the cell to stop dividing when damaged.

Estrogen, Transcription and Cancer (page 113)

1. Estrogen binds to Estrogen receptors and acts as a transcription factor for the transcription of the AID gene.
2. AID causes somatic hypermutation in the DNA of B cells. This causes them to produce novel copies of antibodies which are produced against a myriad (as yet not encountered) antigens.
3. High levels of Estrogen allow for the greater production of AID because more oestrogen is available to initiate the transcription pathway. Normally, levels of Estrogen fluctuate in the body. In men they are relatively lower, but in women the levels rise just before ovulation. There is a corresponding rise in immunity at this time.
4. Continual high levels of Estrogen result in the AID gene being switched on more than is normal, leading to a higher probability of a translocation mutation. This may also explain the link between women on the oral contraceptive pill and their slightly increased risk of cancer. AID's normal role is to cause role-driven (purposed) somatic hypermutation in B cells, but it can also cause other mutations. Specifically, it can cause a translocation mutation (of the CD95/Fas gene, a death receptor gene), preventing normal cell death and inducing tumour formation.

Stem Cells (page 114)

1. (a) They have the ability to self renew while maintaining an undifferentiated state.
   (b) A stem cell has the capacity to differentiate into

specialized cells types (potency).

2. Embryonic stem cells are pluripotent and can form over 220 cell types in the three primary germ layers (ectoderm, endoderm, mesoderm). Their potential for medical application is vast as they can theoretically be used to replace most damaged cell and tissue types. Adult stem cells are multipotent and can divided only into a limited number of cell types, mainly those of the blood, heart and nerves. Their medical applications are more limited than ESC, but because there are fewer ethical issues involved ASC are already being used to treat medical problems such as leukaemia.

3. The main purpose of ASC are to maintain and repair the tissue in which they are found. Some examples are:
   - Hematopoietic stem cells give rise to all the types of blood cells, red blood cells, lymphocytes, leukocytes and platelets.
   - Bone marrow stromal cells (mesenchymal stem cells) give rise to a variety of cell types including bone cells (osteocytes), cartilage cells (chondrocytes), fat cells (adipocytes) and connective tissue cells such as tendons.
   - Neural stem cells in the brain can produce nerve cells (neurons) and the non-neuronal cells astrocytes and oligodendrocytes.
   - Epithelial stem cells that line the digestive tract give rise to absorptive cells, globet cells, Paneth cells and enteroendocrine cells.
   - Skin stem cells in the basal layer of the epidermis form keratinocytes which protect the skin. Follicular stem cells at the base of hair follicles give rise to the hair follicle and epidermal cells.

4. Advantages of using embryonic stem cell cloning in tissue engineering:
   - This technology will provide a disease-free and plentiful supply of (tissue-typed, compatible) cells.
   - Embryonic stem cells have the ability to develop and form all the tissues of the body. Theoretically, there should be no shortage of specific cell types.
   - It will improve the possibility of creating semi-synthetic living organs for use as replacement parts.

Differentiation of Human Cells (page 115)

1. 230 different cell types

2. 50 cell divisions

3. 100 billion cells

4. (Any one of): Skin cells, intestinal epithelial cells, blood (stem) cells.

5. (Any one of): Nerve cells, bone cells, kidney cells.

6. (a) **Germ line** is the cell line that, early in development, becomes differentiated from the somatic cell line and has the potential to form gametes.
   (b) Germ cells will produce gametes (eggs and sperm) and must be essentially unspecialized cells. This is necessary so that none of the genes that are needed to produce the 230 specialized cells in new offspring are turned off before they are needed.

7. (a) A **clone** is a copy of a cell (or complete organism) with a genetic make-up that is identical to the single parent cell it was created from.
   (b) As for 6(b): None of the genes required to produce specialized cells have been turned off.

8. Cancerous cells are cells that have lost control of the regulatory processes that govern the cell's function. Instead they become generalized cells that lose their tissue identity, pulling away from cells around them and undergoing cell division at a rapid rate.

9. At certain stages in the sequence of cell divisions as the embryo grows, some genes get switched on while others get switched off permanently, causing the cells to take on specialized functions.

Stem Cells and Tissue Engineering (page 117)

1 A tissue engineered skin product allows victims to make skin akin to their own and thereby circumventing any rejection problems.

2. (a) The major difficulty with cell transplants is the problem of achieving histocompatibility and preventing an inappropriate immune response to (and rejection of) the cells.
   (b) Poor histocompatibility could be overcome with therapeutic stem cloning using ESC and the recipient's own cells. The recipient's cells provide the nucleus and the ESC provides the cytoplasmic environment; this creates a preembryo containing the patient's own DNA.
   (c) One ethical concern with this is the source of the ESCs, as they must come from the sacrifice of blastocysts (i.e. potentially new embryos).

3. As tissue cultures become larger and more complex diffusion is no longer a suitable method of delivering gases, nutrients, and growth substances to the culture. Suitable mass transport delivery methods must be found to suit the type of tissue culture. For example, a stirred bioreactor for blood cell cultures). Controlling the environmental conditions (e.g. pH, temperature) must also be strictly controlled or the culture will die.

Human Cell Specialization (page 119)

1. (b) **Erythrocyte**:
   <u>Features</u>: Biconcave cell, lacking mitochondria, nucleus, and most internal membranes. Contains the oxygen-transporting pigment, hemoglobin.
   <u>Role</u>: Uptake, transport, and release of oxygen to the tissues. Some transport of CO_2. Lack of organelles creates more space for oxygen transport. Lack of mitochondria prevents oxygen use.
   (c) **Retinal cell**:
   <u>Features</u>: Long, narrow cell with light-sensitive pigment (rhodopsin) embedded in the membranes.
   <u>Role</u>: Detection of light: light causes a structural change in the membranes and leads to a nerve impulse (result is visual perception).
   (d) **Skeletal muscle cell(s)**:
   <u>Features</u>: Cylindrical shape with banded myofibrils. Capable of contraction (shortening).
   <u>Role</u>: Move voluntary muscles acting on skeleton.
   (e) **Intestinal goblet cell (secretory cell)**:
   <u>Features</u>: Flask-shaped cell with basal nucleus and a cell interior filled with mucus globules.
   <u>Role</u>: Secrete mucus to protect the epithelium from abrasion and from the action of digestive enzymes.

(f) **Motor neuron cell**:
Features: Cell body with a long extension (the axon) ending in synaptic bodies. Axon is insulated with a sheath of fatty material (myelin).
Role: Rapid conduction of motor nerve impulses from the spinal cord to effectors (e.g. muscle).

(g) **Spermatocyte**:
Features: Motile, flagellated cell with mitochondria. Nucleus forms a large proportion of the cell.
Role: Male gamete for sexual reproduction. Mitochondria provide the energy for motility.

(h) **Osteocyte**:
Features: Cell with calcium matrix around it. Fingerlike extensions enable the cell to be supplied with nutrients and wastes to be removed.
Role: In early stages, secretes the matrix that will be the structural component of bone. Provides strength.

Plant Cell Specialization (page 120)

1. (b) **Pollen grain**:
Features: Small, lightweight, often with spikes.
Role: Houses male gamete for sexual reproduction.

(c) **Palisade parenchyma cell**:
Features: Column-shaped cell with chloroplasts.
Role: Primary photosynthetic cells of the leaf.

(d) **Epidermal cell**:
Features: Waxy surface on a flat-shaped cell.
Role: Provides a barrier to water loss on leaf.

(e) **Vessel element**:
Features: Rigid remains of a dead cell. No cytoplasm. End walls perforated. Walls are strengthened with lignin fibers.
Role: Rapid conduction of water through the stem. Provides support for stem/trunk.

(f) **Stone cell**:
Features: Very thick lignified cell wall inside the primary cell wall. The cytoplasm is restricted to a small central region of the cell.
Role: Protection of the seed inside the fruit.

(g) **Sieve tube member**:
Features: Long, tube-shaped cell without a nucleus. Cytoplasm continuous with other sieve cells above and below it. Cytoplasmic streaming is evident.
Role: Responsible for translocation of sugars etc.

(h) **Root hair cell:**
Features: Thin cuticle with no waxy layer. High surface area relative to volume.
Role: Facilitates the uptake of water and ions.

Levels of Organization (page 121)

1. **Animals**
(a) **Molecular**: Adrenaline, collagen, DNA, phospholipid
(b) **Organelles**: Lysosome, ribosomes
(c) **Cells**: Leukocyte, mast cell, neuron, Schwann cell
(d) **Tissues**: Blood, bone, cardiac muscle, cartilage, squamous epithelium
(e) **Organs**: Brain, heart, spleen
(f) **Organ system**: Nervous system, reproductive system

2. **Plants**
(a) **Molecular**: Pectin, cellulose, DNA, phospholipid
(b) **Organelles**: Chloroplasts, ribosomes
(c) **Cells**: Companion cells, epidermal cell, fibers, tracheid
(d) **Tissues**: Collenchyma*, mesophyll, parenchyma*, phloem, sclerenchyma
(e) **Organs**: Flowers, leaf, roots

* **Note**: Parenchyma and collenchyma are simple tissues comprising only one type of cell (parenchyma and collenchyma cells respectively). Simple plant tissues are usually identified by cell name alone.

Animal Tissues (page 122)

1. The organization of cells into specialized tissues allows the tissues to perform particular functions. This improves efficiency of function because different tasks can be shared amongst specialized cells. Energy is saved in not maintaining non-essential organelles in cells that do not require them.

2. (a) **Epithelial tissues**: Single or multiple layers of simple cells forming the lining of internal and external body surfaces. Cells rest on a basement membrane of fibers and collagen and may be specialized. **Note**: epithelial cells may be variously shaped: squamous (flat), cuboidal, columnar etc.
(b) **Nervous tissue**: Tissue comprising densely packed nerve cells specialized for transmitting electro-chemical impulses. Nerve cells may be associated with supportive cells (e.g. Schwann cells), connective tissue, and blood vessels.
(c) **Muscle tissue**: Dense tissue comprising highly specialized contractile cells called fibers held together by connective tissues.
(d) **Connective tissues**: Supporting tissue of the body, comprising cells widely dispersed in a semi-fluid matrix (or fluid in the case of blood and lymph).

3. (a) Muscle tissue is made up of long muscle fiber cells made up of myofibrils. The myofibrils are made up of contractile proteins actin and myosin, which cause the muscle fibers to contract when stimulated. The contraction results in movement of the organism itself (locomotion) or movement of an internal organ.
(b) Nervous tissue comprises two main tissue types: neurons which transmit nerve signals and glial cells which provide support to the neurons. Neurons have several protrusions (dendrites or axons) from their cell body which allow conduction of nerves impulses to target cells.

Plant Tissues (page 123)

1. **Collenchyma**
Cell type(s): collenchyma cells
Role: provides flexible support.

Sclerenchyma
Cell type(s): sclerenchyma cells
Role: provides rigid, hard support.

Root Endodermis
Cell type(s): endodermal cells
Role: Provides selective barrier regulating the passage of substances from the soil to the vascular tissue.

Pericycle
Cell type(s): parenchyma cells
Role: Production of branch roots, synthesis and transport of alkaloids.

Leaf mesophyll
Cell type(s): spongy mesophyll, palisade mesophyll

Role: Main photosynthesis site in the plant.

Xylem
Cell type(s): tracheids, vessel members, fibers, paraenchyma cells
Role: Conducts water and dissolved minerals in vascular plants.

Phloem
Cell type(s): sieve-tube members, companion cells, parenchyma, fibers, sclereids
Role: transport of dissolved organic material (including sugars) within vascular plants.

Epidermis
Cell type(s): epidermal cells, guard cells, subsidiary cells, and epidermal hairs (trichomes).
Role: Protection against water loss, regulation of gas exchange, secretion, water and mineral absorption.

Root Cell Development (page 124)

1. (a) Cells specialize to take on specific functions.
   (b) Cells are becoming longer and/or larger.
   (c) Cells are dividing by mitosis.
2. (a) Late anaphase; chromatids are being pulled apart and are at opposite poles.
   (b) Telophase; there are two new nuclei formed and the cell plate is visible.
   (c) 25 of 250 cells were in mitosis, therefore mitosis occupies 25/250 or one tenth of the cell cycle.
3. The **cambium layer** of cells (lying under the bark between the outer phloem layer of cells and the inner xylem layer of cells). **Note**: Cells dividing from each side of this layer specialize to form new phloem on the outside and new xylem on the inside.

KEY TERMS: Mix and Match (page 125)

Active transport (F), Amphipathic (BB), Cellular differentiation (W), Concentration gradient (V), Cytokinesis (X), Diffusion (O), Emergent property (I), Endocytosis (U), Exocytosis (Y), Facilitated diffusion (C), Fluid mosaic model (B), Gamete (CC), Glycolipids (K), Glycoproteins (AA), Hypertonic (Q), Hypotonic (G), Ion pump (T), Isotonic (L), Mitosis (N), Osmosis (A), Oncogene (R), Partially permeable membrane (Z), Passive transport (P), Phagocyctosis (H), Plasma membrane (S), Plasmolysis (E), Surface area: volume ratio (M), Turgor (J), Water potential (D).

The Role of ATP in Cells (page 127)

1. In the presence of the enzyme ATPase, ATP is hydrolyzed to produce ADP plus a free phosphate, releasing energy in the process.
2. Glucose
3. Cellular respiration; strictly oxidative phosphorylation
4. Solar energy
5. Food (gaining nutrient from plants and other animals)
6. Like a rechargeable battery, the ADP/ATP system toggles between a high energy state and a low energy. The addition of a phosphate to ADP recharges the molecule so that it can be used for cellular work.
7. **PHOTOSYNTHESIS**
   Starting materials: carbon dioxide, water (as a source of hydrogens), in the presence of light and chlorophyll.
   Waste products: oxygen, water.
   Role of hydrogen carriers: NADP: Carries hydrogen between light dependent and light independent phases (where the hydrogen is incorporated into sugars).
   Role of ATP: Produced in the light dependent phase and used in the light independent phase to make sugars from carbon dioxide and hydrogen.
   Overall biological role: Uses light energy to fix carbon into organic molecules which become part of the energy available in food chains.

   CELLULAR RESPIRATION
   Starting materials: organic molecules (ultimately glucose), oxygen.
   Waste products: carbon dioxide, water.
   Role of hydrogen carriers: NAD: Carries hydrogens to the electron transport system where their transfer between carriers is coupled to ATP production.
   Role of ATP: A small amount of ATP is used initially to produce pyruvate from glucose. Produced in glycolysis, Krebs cycle, and the ETS.
   Overall biological role: The process by which organisms break down energy rich molecules to release energy in a usable form (ATP).

Measuring Respiration (page 129)

1. (a) RQ at 20°C/48 h = 1.97 ÷ 2.82 = **0.7** (0.698)
   (b) RQ at 20°C/1 h = 2.82 ÷ 2.82 = 1. After 2 days without feeding the cricket was metabolizing fats for energy. Shortly (1 hour) after feeding it was metabolizing only carbohydrate (RQ = 1).
2. (a) RQ of two seedlings during early germination:

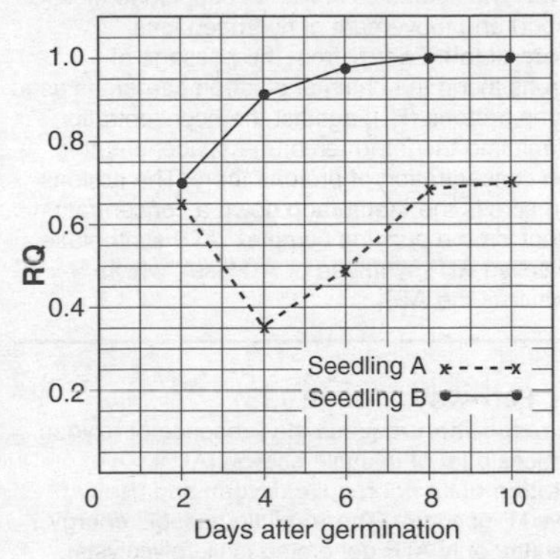

   (b) Both seedlings began their germination metabolizing primarily fats for energy. However, while seedling A continued to metabolize mainly fats (with some synthesis of carbohydrate and organic acids), throughout the 10 day period, seedling B rapidly moved to metabolizing carbohydrate alone. **Note**: The value of 0.91 (seedling B) may have been the result of protein metabolism alone or (more likely) respiration of a mix of fat and glucose.

Cellular Respiration (page 130)

1. (a) Glycolysis: cytoplasm
   (b) Krebs cycle: matrix of mitochondria
   (c) Electron transport chain: cristae (inner membrane surface) of mitochondria.
2. (a) Substrate-level phosphorylation: Formation of ATP by the transfer a phosphate group from a substrate (e.g. a phosphorylated 6C sugar) to ADP directly, with no involvement of an electron transport chain.
   (b) Oxidative phosphorylation: The process by which glucose is oxidized in a series of redox reactions and the energy released in the electron transfers is coupled to ATP synthesis.

The Biochemistry of Respiration (page 131)

1. (a) 6 carbon atoms
   (b) 3 carbon atoms (glucose split into two)
   (c) 2 carbon atoms (1 carbon lost as CO_2)
   (d) 6 carbon atoms (2-carbon acetyl added to 4 carbon)
   (e) 5 carbon atoms (1 carbon lost as CO_2)
   (f) 4 carbon atoms (1 carbon lost as CO_2)
2. (a) Glycolysis: 2 ATPs
   (b) Krebs cycle: 2 ATPs
   (c) Electron transport chain: 34 ATPs
   (d) Total produced: 38 ATPs
3. Carbon atoms are released as carbon dioxide (CO_2) gas and breathed out through gas exchange surfaces.
4. (a) Hydrogen atoms supply energy in the form of high energy electrons. **Note**: These are passed along the respiratory chain, losing energy as they go. The energy released is used to generate ATP.
   (b) Oxygen is the final electron acceptor at the end of the respiratory chain.
5. (a) ATP is generated by **chemiosmosis**.
   (b) **In brief**: The synthesis of ATP is coupled to electron transport and movement of hydrogen ions.
   In more detail: Energy from the passage of electrons along the chain of electron carriers is used to pump protons (H^+), against their concentration gradient, into the intermembrane space, creating a high concentration of protons there. The protons return across the membrane down a concentration gradient via the enzyme complex, ATP synthetase (also called ATP synthase or ATPase), which synthesizes the ATP.

Anaerobic Pathways (page 133)

1. **Aerobic respiration** requires the presence of oxygen and produces a lot of useable energy (ATP).
   Fermentation does not require oxygen and uses an alternative H^+ acceptor. There is little useable energy produced (the only ATP generated is via glycolysis).
2. (a) $2 \div 38 \times 100 = 5.3\%$ efficiency
   (b) Only a small amount of the energy of a glucose molecule is released in anaerobic respiration. The remainder stays locked up in the molecule.
3. The build up of toxic products (ethanol or lactate) inhibits further metabolic activity.

Photosynthesis (page 134)

1. (a) Grana: Membranes that contain chlorophyll and are the site of the light dependent phase of photosynthesis. The biochemical process involves energy capture via photosystems I and II.
   (b) Stroma: The liquid interior of the chloroplast in which the light independent phase takes place. The biochemical process involves carbon fixation via the Calvin cycle.
2. (a) Carbon dioxide: Comes from the air (via stomata) and provides carbon and oxygen as raw materials for the production of glucose. Some oxygen molecules contribute to the production of H_2O
   (b) Oxygen: Comes from CO_2 gas (via stomata) and H_2O (via the roots and vascular system). The oxygen from the CO_2 is incorporated into glucose and H_2O. The oxygen from the water is given off as free oxygen (a waste product).
   (c) Hydrogen: Comes from water (via the roots and vascular system) obtained from the soil. This hydrogen is incorporated into glucose and H_2O.
   Note: The clarify: isotope studies show that the carbon and oxygen in the carbohydrate comes from CO_2, while the free oxygen comes from H_2O.
3. Scientists use isotopes to tag molecules taken up by plants and analyse the products produced as a result of photosynthesis. For example, the use of oxygen isotopes can identify the origin of the free oxygen released in photosynthesis:
   $CO_2 + 2H_2{}^{18}O \rightarrow [CH_2O] + {}^{18}O_2 + H_2O$
4. Glucose is used as the fuel for cellular respiration, or used to construct cellulose, starch, or disaccharide molecules (e.g. fructose). Oxygen is required for aerobic respiration and water is recycled and even reused for photosynthesis.

Pigments and Light Absorption (page 135)

1. The absorption spectrum of a pigment is that wavelength of the light spectrum absorbed by a pigment, e.g. chlorophyll absorbs red and blue light and appears green. Represented graphically, the absorption spectrum shows the relative amounts of light absorbed at different wavelengths.
2. Accessory pigments absorb light wavelengths that chlorophyll *a* cannot absorb, and they pass their energy on to chlorophyll *a*. This broadens the action spectrum over which chlorophyll *a* can fuel photosynthesis.

Photosynthetic Rate (page 136)

1. (a) Photosynthetic rate increases rapidly then levels off.
   (b) Up to a certain light intensity more light is available to the chlorophyll so the rate increases. When all the chlorophyll molecules are activated (saturated) by the light, more light has no further effect.
2. (a) Increased temperature increases the photosynthetic rate, but this effect is not marked at low CO_2.
   (b) At higher temperature biochemical reactions occur more rapidly. At low CO_2 levels, rate is determined more by the CO_2 (raw material) available (rates are low regardless of temperature).
3. The photosynthetic rate is determined by the rate at

which CO_2 enters the leaf. When this declines because of low atmospheric levels, so does photosynthetic rate.

4. (a) By changing only one factor at a time (either temperature or CO_2 level) it is possible to assess the effects of each one separately.
   (b) CO_2 has the greatest effect of these two variables.
   (c) At low levels of CO_2, increase in temperature has little effect (the rate of CO_2 entry into the leaf is the greatest determinant of photosynthetic rate).

The Biochemistry of Photosynthesis (page 137)

1. **NADP**: Carries H_2 from the light dependent phase to the light independent reactions.
2. **Chlorophyll**: These pigment molecules trap light energy and produce high energy electrons. These are used to make ATP and NADPH. The chlorophyll molecules also split water, releasing H^+ for use in the light independent reactions and liberating free O_2.
3. (a) 1C (b) 3C (c) 3C (d) 3C (e) 3C (f) 3C (g) 5C (h) 5C
4. Six turns.
5. (a) **Light dependent (D) phase**: Takes place in the grana (thylakoid membranes) of the chloroplast and requires light energy to proceed. The light dependent phase generates ATP and reducing power in the form of NADPH. The electrons and hydrogen ions come from the splitting of water.
   (b) **Calvin cycle**: Takes place in the stroma of the chloroplast and does not require light energy to proceed. The Calvin cycle uses ATP and NADPH from the light dependent phase, for the step-wise reduction of carbon dioxide to glucose.
6. Carbon and oxygen from carbon dioxide gas (via stomata). Hydrogen from water (via the roots and vascular system) obtained from the soil. **Note**: It has been shown through oxygen isotope studies that the free oxygen produced as a result of photosynthesis comes from water and the oxygen in the carbohydrate comes from carbon dioxide.
7. The ATP synthesis is coupled to electron transport. When the light strikes the chlorophyll molecules, high energy electrons are released by the chlorophyll molecules. The energy lost when the electrons are passed through a series of electron carriers is used to bond a phosphate to ADP to make ATP.

   Note: ATP is generated (in photosynthesis and cellular respiration) by **chemiosmosis**. As the electron carriers pick up the electrons, protons (H^+) pass into the space inside the thylakoid, creating a high concentration of protons there. The protons return across the thylakoid membrane down a concentration gradient via the enzyme complex, ATP synthetase that synthesizes the ATP (also called ATP synthase or ATPase).
8. (a) **Non-cyclic (photo)phosphorylation**: Generation of ATP using light energy during photosynthesis. The electrons lost during this process are replaced by the splitting of water.
   (b) The term non-cyclic **photo**phosphorylation is also (commonly) used because it indicates that the energy for the phosphorylation is coming from light.
   (c) In oxidative phosphorylation, the energy for the phosphorylation comes from the oxidation of glucose (the transfer of electrons released in the oxidation is coupled to ATP synthesis).
9. (a) In cyclic photophosphorylation, the electrons lost from photosystem II are replaced by those from photosystem I rather than from the splitting of water. ATP is generated in this process, but not NADPH
   (b) The non-cyclic path produces ATP and NADH in roughly equal quantities but the Calvin cycle uses more ATP than NADPH. The cyclic pathway of electron flow makes up the difference.

Photosynthesis in C_4 Plants (page 139)

1. At high temperatures, C_4 plants are able to fix carbon dioxide in the palisade mesophyll (even at low concentrations), and transport malate into the bundle sheath to enhance carbon dioxide levels for the Calvin cycle. In this way, they can achieve higher rates of photosynthesis and a greater yield of photosynthetic product than C_3 plants. (**Note**: High levels of CO_2 inhibit photorespiration and improve the efficiency of RuBisCo function. Photorespiration is a major cause of lower photosynthetic activity in C_3 plants).
2. Oxygen is a competitive inhibitor of the enzyme RuBisCo (ribulose bisphosphate carboxylase). **Note:** Catalyses the reaction between RuBP (ribulose bisphosphate) and CO_2. Keeping oxygen low in the bundle sheath prevents RuBisCo inhibition and enhances photosynthetic rate.
3. Most C_4 species live in hot environments where they can outcompete C_3 plants because of their higher rates of photosynthesis. **Note**: The hot climates are required because many of the enzymes of the C_4 cycle have optimum temperatures above 25°C. In addition, C_4 plants also require a high rate of respiration to generate an excess of the carbon dioxide acceptor, phosphoenol pyruvate (PEP).
4. Higher ambient CO_2 levels enhance CO_2 levels in the mesophyll, preventing the photorespiration that normally occurs as a result of RuBisCo inhibition. C_4 plants do this anyway by boosting CO_2 levels in the bundle sheath through the C_4 pathway. Increasing ambient CO_2 therefore produces no further increase in photosynthetic rate in C_4 plants.

KEY TERMS: Mix and Match (page 140)

Absorption Spectrum (F), Accessory Pigment (BB), Acetyl Coenzyme A (W), Action Spectrum (V), ATP (X), Calvin Cycle (O), Cellular Respiration (I), Chemiosmosis (U), Chloroplast (Y), Chlorophyll (C), Cristae (B), Electron Transport Chain (D), Fermentation (K), Glycolysis (AA), Hatch and Slack Pathway (Q), Kreb's Cycle (G), Lactic acid (T), Matrix (L), Mitochondrion (N), Non-Cyclic Photophosphorylation (A), Oxidative Phosphorylation (R), Photolysis (Z), Photosynthesis (P), Respiratory Quotient (H), Ribulose Bisphosphate (S), Stroma (E), Substrate level Phosphorylation (M), Thylakoid Discs (J).

Meiosis (page 143)

1. In the first division of meiosis, homologous pairs of chromosomes pair to form bivalents. Segments of chromosome may be exchanged in crossing over and the homologues then separate (are pulled apart). This division reduces the number of chromosomes in the intermediate cells, so that only one chromosome from each homologous pair is present.
2. In the second division of meiosis, chromatids separate (are pulled apart), but the number of chromosomes stays the same. This is more or less a 'mitotic' division.
3. Mitosis involves a division of the chromatids into two new daughter cells thus maintaining the original number of chromosomes in the parent cell. Meiosis involves a division of the homologous pairs of chromosomes into two intermediary daughter cells thus reducing the diploid number by half. The second stage of meiosis is similar to a mitotic division, but the haploid number is maintained because the chromatids separate.
4. **A** shows metaphase of meiosis I; the homologous pairs of chromosomes are lined up on the cell equator. **B** shows metaphase of meiosis II; the individual chromatids are about to separate.

Crossing Over (page 145)

1. Unexpected combinations of alleles for genes will occur that would not normally be present in gametes.
2. Crossing over provides one source of genetic variation amongst individuals in a population. This is important for providing the raw material upon which natural selection acts.

Mitosis vs Meiosis (page 146)

1. Mitosis involves a division of the chromatids into two new daughter cells thus maintaining the original number of chromosomes in the parent cell. Meiosis involves a division of the homologous pairs of chromosomes into two intermediary daughter cells thus reducing the diploid number by half. The second stage of meiosis is similar to a mitotic division, but the haploid number is maintained because the chromatids separate.
2. The first division is a reduction division, halving the number of chromosomes. The second division is a 'mitotic' type division, the chromatids are separated but the number of chromosomes remain the same.
3. The processes of recombination and independent assortment of the chromosomes during meiosis introduces genetic diversity; the alleles are mixed up into different combinations in the offspring.

Linkage (page 147)

1. **Linkage** refers to the situation where genes are located on the same chromosome. As a result, they tend to be inherited together as a unit.
2. (a) AaBb, Aabb, aaBb, aabb
   (b) F_1 genotype: all CucuEbeb (heterozygotes)
   F_1 phenotype: all wild type (straight wing, gray body).
3. Gene linkage reduces the amount of variation because the linked genes are inherited together and fewer genetic combinations of their alleles are possible.

Recombination (page 148)

1. **Recombination** refers to the exchange of alleles between homologous chromosomes as a result of crossing over. It produces new associations of alleles in the offspring.
2. Recombination increases the amount of genetic variation because parental linkage groups separate and new associations of alleles are formed in the offspring. The offspring show new combinations of characters that are unlike the parental type.
3. A greater than 50% recombination frequency indicates that there is independent assortment (the genes must be on separate chromosomes).

The Advantages of Sex (page 149)

1. Sexual reproduction produces genetic variability in the offspring - different individuals have different allele combinations. This provides the flexibility to adapt to changing conditions by "selecting" adaptive genotypes (different allele combinations will have different reproductive success in different environments).
2. (a) Species that reproduce asexually can do so at a much greater rate than sexually reproducing organisms (time is not spent in mating, gamete production etc.). Provided the environment is stable, a well adapted clone can saturate the environment and outcompete sexually reproducing organisms (assuming some niche overlap).
   (b) Asexual organisms would not be favored in a changeable environment, as they lack the genetic variability from which to select adaptive genotypes.
3. In sexually reproducing organisms, there is the opportunity for favorable mutations to come together in an individual. In asexual organisms, clones with a favorable mutation compete against other (well adapted) clones.
4. Species that alternate between sexual and asexual reproduction combine the advantages of being able to saturate the environment with well adapted clones (asexual) with the production (through meiosis and fertilisation) of individuals with different allele combinations. The genetic variability provides the opportunity to produce new, well adapted genotypes.

Genomes (page 150)

1. The **genome** of an organism is a complete haploid set of all chromosomes (i.e. all the genetic material carried by a single representative of each of all chromosome pairs).
2. (a) 5375 bases (b) 5.375 kb (c) 0.005375 Mb
3. 1542 bases
4. It is a comparatively small genome, others having 10 to 40 times as much genetic material (e.g. 48.6-190 kb).

Eukaryote Chromosome Structure (page 151)

1. (a) **DNA**: A long, complex nucleic acid molecule found in the chromosomes of nearly all organisms (some viruses have RNA instead). Provides the genetic instructions (genes) for the production of proteins and other gene products (e.g. RNAs).
   (b) **Chromatin**: Chromosomal material consisting of DNA, RNA, and histone and non-histone proteins. The term is used in reference to chromosomes in the non-condensed state.
   (c) **Histone**: Simple proteins that bind to DNA and help it to coil up during cell division. Histones are also involved in regulating DNA function in some way.
   (d) **Centromere**: A bump or constriction along the length of a chromosome to which spindle fibers attach during cell division. The centromere binds two chromatids together.
   (e) **Chromatid**: One of a pair of duplicated chromosomes produced prior to cell division, joined at the centromere. The terms chromatid and chromosome distinguish duplicated chromosomes before and after division of the centromere.

2. The chromatin (DNA and associated proteins) combine to coil up the DNA into a "super coiled" arrangement. The coiling of the DNA occurs at several levels. The DNA molecule is wrapped around bead-like cores of (8) histone proteins (called nucleosomes), which are separated from each other by linker DNA sequences of about 50 bp. The histones (H1) are responsible for pulling nucleosomes together to form a 30 nm fiber. The chromatin fiber is then folded and wrapped so that it is held in a tight configuration. The different levels of coiling enables a huge amount of DNA to be packed, without tangling, into a very small space in a well organized, orderly fashion.

Karyotypes (page 153)

1. A **karyotype** is the chromosome complement of a cell or organism, characterized by the number, size, shape, and centromere position of the chromosomes. It can provide information on gender and chromosomal abnormalities, such as Down syndrome.

2. **Autosomes** are the non-sex chromosomes that occur as 'matching' homologous pairs and are not involved in determining the sex of the organism. In contrast, the **sex chromosomes** (also called heterosomes) are the pair of chromosomes (XX in female humans and XY in male humans) that determine gender.

3. Number the chromosomes: See below.

4. Circle the sex chromosomes: See below.

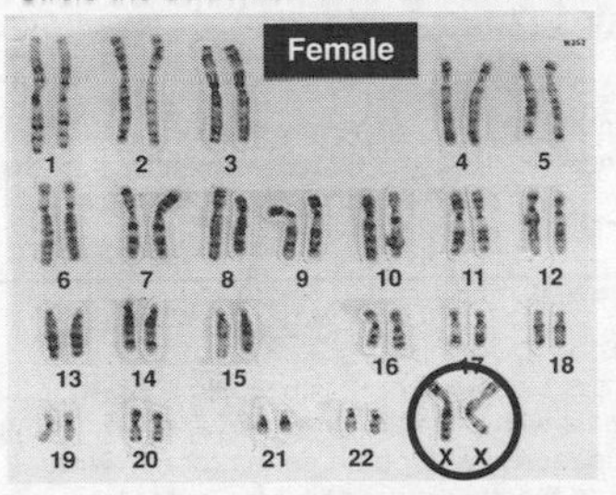

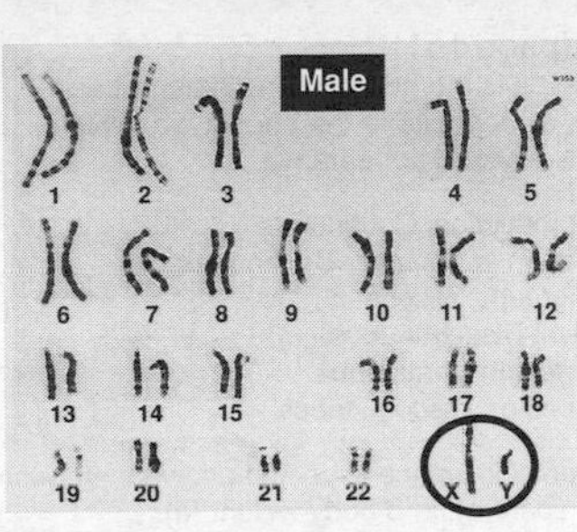

5. (a) Female autosomes: 44, sex chromosomes: XX
   (b) Male autosomes: 44, sex chromosomes: XY

6. (a) 46 (b) 23

Human Karyotype Exercise (page 155)

By studying the distinguishing characteristics of the chromosomes, you should be able to arrange the them in their correct sequence on the karyotype record sheet.

1. The karyotype should be organized as for the "Typical layout of a human karyotype" on page 169, but it has **one extra chromosome number 21 (trisomy 21),** and is **male** (XY) rather than female (XX).

2. (a) Sex: male
   (b) Abnormal
   (c) 45 + XY (trisomy 21 or Down syndrome)

KEY TERMS: Mix and Match (page 158)

Allele (F), Amniocentesis (B), Anaphase 1 (W), Autosome (V), Bivalent (X), Chorionic villus sampling (O), Chromatid (I), Chromosome (U), Crossing over (Y), Crossover frequency (C), Diploid (2N) (BB), Fertilization (D), Gamete (K), Haploid (1N) (A), Histone protein (Q), Independent assortment (G), Interphase (T), Karyotype (CC), Linkage (N), Maternal chromosome (AA), Meiosis (R), Metaphase 1 (Z), Paternal chromosome (P), Prophase 1 (H), Recombination (S), Sex chromosomes (E), Somatic cell (M), Synapsis (L), Telophase1 (J).

A Gene That Can Tell Your Future? (page 160)

1. The physical effects of Huntington's disease are shaking of hands and/or limbs and an awkward gait. More severe effects include the loss of muscle control and mental function leading to dementia.

2. The mHTT gene was discovered using information from the family history of 10,000 people. Using a probe called G8, a map of the 4th chromosome was built up and each gene sequenced. The mHTT gene was shown to be one with a trinucleotide repeat expansion.

3. HD is caused by a trinucleotide repeat expansion of the sequence CAG on the 4th chromosome. Repeats of over 35 cause the disease and the greater the number of repeats the more severe the disease Because of the instability of the mHTT gene the number of repeats and severity of the disease tends to increase over generations.

Variation (page 161)

1. The genotype is the genetic constitution of an organism, as opposed to the phenotype, which is the organism's physical appearance.

2. (a) Wool production: Continuous
   (b) Hand span in humans: Continuous
   (c) Blood groups: Discontinuous
   (d) Albinism: Discontinuous
   (e) Body weight: Continuous
   (f) Flower color: Discontinuous

3. Environmental influence expected on: wool production (a), Hand span (b), and body weight (e).

4. Wind speed, temperature, air density, water availability.

5. The acidity of the soil. In alkaline soils they are pink or red-purple, in acid soil the flowers are blue. The color is due to the presence or absence of aluminium compounds in the flowers. Plants can only access aluminium compounds when soil pH is low. When aluminium is present within the plant, the flowers are blue. If absent they are pink.

6. (a) A cline is a continuous, or nearly continuous, gradation in a phenotypic character within a species, associated with a change in an environmental variable such as temperature.
   (b) Plant species A: The observed phenotype (prostrate) is not due to genetic factors, but to the effect of climate on growth. In the absence of a harsh environment, the plant reverts to its normal growing habit
   Plant species B: The low growing (prostrate) phenotype of this species is controlled by genes (not by environmental factors).
   (c) Plant A

Alleles (page 163)

1. (a) Heterozygous: Each of the homologous chromosomes contains a **different** allele for the gene (one dominant and one recessive).
   (b) Homozygous dominant: Each of the homologous chromosomes contains an identical dominant allele.
   (c) Homozygous recessive: Each of the homologous chromosomes contains an identical recessive allele.

2. (a) **Aa** (b) **AA** (c) **aa**

3. Each chromosome of a homologous pair comes from a different parent: one of maternal origin, one of paternal origin (they originated from the egg and the sperm that formed the zygote). They contain the same sequence of genes for the same traits, but the versions of the genes (alleles) on each chromosome may differ.

4. **Alleles** are different versions of the same gene that code for the same trait. Different alleles provide phenotypic variation for the expression of a gene. There are often two alleles for a gene, one dominant and one recessive. In this case, the dominant allele will be expressed in the phenotype. Sometimes alleles for a gene can be equally dominant, in which case, both alleles will be expressed in the phenotype. Where three or more alleles for a gene exist (multiple alleles), there is more phenotypic variation in the population (for that trait) than would be the case with just two alleles.

Mendel's Pea Plant Experiments (page 164)

1. and 2. (see table below):

	Dominant	Recessive	Ratio
Seed color	Yellow	Green	3.01 : 1
Pod color	Green	Yellow	2.82 : 1
Flower position	Axial	Terminal	3.14 : 1
Pod shape	Constricted	Inflated	2.95 : 1
Stem length	Tall	Dwarf	2.84 : 1

3. (a) Seed shape (2.96:1), seed color (3.01:1), and pod shape (2.95:1).
   (b) Considering all the traits, larger sample sizes generally produced ratios closer to the predicted theoretical ratio. Smaller samples are more likely to produce results that deviate from the theoretical ideal, because they are affected more by the randomness of meiosis and fertilization.

Mendel's Laws of Inheritance (page 165)

1. Particulate inheritance: Inherited characteristics are transmitted by discrete entities (genes) which themselves remain unchanged from generation to generation. **Note**: Flower color is controlled by two alleles, a dominant purple one and a recessive white one. All offspring receive one of each of the alleles, but only the dominant one is expressed. In subsequent offspring, recessive alleles may be provided by each of the gametes to produce white flowers.

2. **Note**: During meiosis, the two alleles for a gene will separate into different gametes, and subsequently into different offspring. Normally both alleles cannot end up in the same offspring. Occasionally, faulty meiosis can occur, resulting in aneuploidy or polyploidy.
   (a) Aa (b) A, A, a, a (c) 2 kinds: A and a

3. **Note**: During meiosis, all combinations of alleles are distributed to gametes with equal probability. The pair of alleles for each gene are sorted independently of those for all other genes. Genes that are linked on the same chromosome tend to be inherited together.
   (a) AB and ab (b)4 kinds: AB, Ab, aB, ab

Basic Genetic Crosses (page 166)

1.

	YR	Yr	yR	yr
YR	YYRR	YYRr	YyRR	YyRr
Yr	YYRr	YYrr	YyRr	Yyrr
yR	YyRR	YyRr	yyRR	yyRr
yr	YyRr	Yyrr	yyRr	yyrr

2. Yellow-round: 9/16 Yellow-wrinkled: 3/16
   Green-round: 3/16 Green-wrinkled: 1/16

3. Ratio: 9 : 3 : 3 : 1

The Test Cross (page 167)

1. To perform the test cross here you would have to cross the male with a female (brown eye, ebony body) homozygous recessive for both genes (bbee). **Note**: Depending on the male's genotype the outcomes of the

test cross would produce offspring that were:
- All wild type (brown body red eye)
- Half wild type, and half red eye, ebony body
- Half wild type, and half brown eye, brown body
- 25% wild type, 25% red eye ebony, 25% brown eye, brown body, and 25% brown eye, ebony.

Students could present this in Punnett squares.

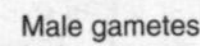

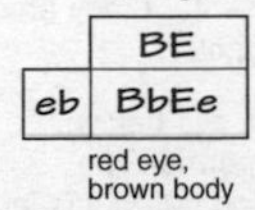

Female gametes

	BE
eb	BbEe
	red eye, brown body

	BE	Be
eb	BbEe	Bbee
	red eye, brown body	red eye, ebony body

	BE	bE
eb	BbEe	bbEe
	red eye, brown body	brown eye, brown body

	BE	Be	bE	be
eb	BbEe	Bbee	bbEe	bbee
	red eye, brown body	red eye, ebony body	brown eye, brown body	brown eye, ebony body

2. A wild type male (brown or normal body and red eyes) could have one of four genotypes: BEBE, BEbE, BEBe, or BEbe

3. (a) 50% are wild type and 50% are red eye, ebony so the male must BEBe.
   (b) This is the only genotype that produces this ratio of phenotypes in the offspring (as per the second Punnett square above)

Monohybrid Cross (page 168)

	Genotype	Phenotype
Cross 2	50% BB 50% Bb	100% black
Cross 3	25% BB 50% B b 25% bb	75% black 25% white
Cross 4	100% BB	100% black
Cross 5	50% Bb 50% bb	50% black 50% white
Cross 6	100% bb	100% white

Dominance of Alleles (page 169)

1. (a) Incomplete dominance: Neither allele is dominant (neither can mask the other).
   (b) Codominance: Two or more alleles are dominant over any recessive alleles; both are fully expressed.

2. (a) Incomplete dominance: Heterozygote's phenotype is intermediate between homozygous parents.
   (b) Codominance: Heterozygotes have a phenotype that is different from either homozygous parent.

3. Ratio: **1 : 2 : 1** (1 red : 2 roan : 1 white).

Male gametes / Female gametes

	C^R	C^W
C^R	C^RC^R	C^RC^W
C^W	C^RC^W	C^WC^W

4. Parents: white and pink. Will produce 50% pink and 50% white offspring.

Male gametes / Female gametes

	C^W	C^W
C^R	C^RC^W	C^RC^W
C^W	C^WC^W	C^WC^W

5. (a) Diagram labels:

	White bull	**Roan cow**
Parent genotype:	C^WC^W	C^RC^W
Gametes:	C^W, C^W	C^R, C^W
Calf genotypes:	C^RC^W, C^WC^W	C^RC^W, C^WC^W
Phenotypes:	Roan, white	Roan, white

   (b) Phenotype ratio: 50% roan, 50% white (1:1)
   (c) By choosing only the roan calves (male and female) to breed from. Initial offspring from roan parents should include all phenotypes: white, roan and red. By selecting just the red offspring from this generation it would be possible to breed a pure herd of red cattle. The unwanted phenotypes must be prevented from breeding (e.g. by castration).

6. (a) Diagram labels:

	Unknown bull	**Roan cow**
Parent Genotype:	C^RC^R	C^RC^W
Gametes:	C^R,C^R	C^R,C^W
Calf Genotypes:	C^RC^R, C^RC^W, C^RC^R, C^RC^W	
Phenotypes:	Red, Roan, Red, Roan	

   (b) Unknown bull: **red** bull

7. (a) Diagram labels:

	Pink	**Red**
Parent Genotype:	C^RC^W	C^RC^R
Gametes:	C^n,C^W	C^n,C^n
Offspring:	C^RC^R, C^RC^R, C^RC^W, C^RC^W	
Phenotypes:	Red, Red, Pink, Pink	

   (b) Phenotype ratio: 50% red, 50% pink (1:1)

Multiple Alleles in Blood Groups (page 171)

Note: In this activity, any reference to the I alleles has been removed on the advice of researchers involved in the latest work in this area. References to the I gene are now considered to be incorrect and misleading.

1. Blood group table:

Blood group **B**	BB, BO
Blood group **AB**	AB

2.

Cross 2	**Group O**	**Group O**
Gametes:	O, O	O, O
Children's genotypes:	OO, OO	OO, OO
Blood groups:	O, O	O, O

Cross 3	**Group AB**	**Group A**
Gametes:	A, B	A, O
Children's genotypes:	AA, AO	BA, BO
Blood groups:	A, A	AB, B

Cross 4	**Group A**	**Group B**
Gametes:	A, A	B, O
Children's genotypes:	AB, AO	AB, AO
Blood groups:	AB, A	AB, A

Cross 5	**Group A**	**Group O**
Gametes:	A, O	O, O
Children's genotypes:	AO, AO	OO, OO
Blood groups:	A, A	O, O

Cross 6	**Group B**	**Group O**
Gametes:	B, O	O, O
Children's genotypes:	BO, BO	OO, OO
Blood groups:	B, B	O, O

Note: Depending on which gamete circle each symbol of a gene is placed, it is possible to have the answers arranged differently. There are of course many more crossover combinations possible.

3. (a) Parent genotypes:

Parent genotype:	AO	OO
Gametes:	A, O	O, O
Children's genotypes:	AO, AO	OO, OO
Blood groups:	A, A	O, O

(b) 50% (c) 50% (d) 0%

4. (a) Possible parent genotypes: Mother assumed to be heterozygous to get maximum variation in gametes (homozygous would also work).

Phenotypes	**Group A**	**Group O**
Genotypes:	AO	OO
Gametes:	A, O	O, O

Child's genotype would have to be AO or OO

(b) Therefore the only possible offspring from this couple would have been children with group **A** or group **O**. The man making the claim could not have been the father of the child.

5. (a) **A** or **B** (b) **AB, A, B** or **O**

Dihybrid Cross (page 173)

Cross N°. 1: Q1-3 integrated

Genotypes and ratios:	BbLL	2	bbLl	4
	BbLl	4	Bbll	2
	bbLL	2	bbll	2
Phenotypes and ratios:	6 black/short		6 white/short	
	2 black/long		2 white/long	

Cross N°. 2: Q1-4 integrated

Gametes: White parent: all bL
Black parent: Bl, Bl, bl, bl

Genotypes and ratios: BbLl 8 bbLl 8 — 1:1 ratio

Phenotypes and ratios: 8 black/short 8 white/short — 1:1 ratio

Dihybrid Cross with Linkage (page 175)

1. Crossover value (COV) for the offspring of the test cross: For crossover value, use the formula:
No. of recombinants ÷ total no. of offspring X 100

(21 + 27) ÷ (123 + 129 + 21 + 27) X 100
= (48 ÷ 300) X 100 = **16%**

Lethal Alleles (page 176)

1. **Recessive** lethal alleles are lethal only when they occur in the homozygous recessive state, whereas **dominant** lethal alleles are either lethal even when only one copy of the allele is present, or they produce a measurable effect in the heterozygote.

2. (a) MM^L and MM (M^LM^L is lethal, see below).
(b) Phenotype ratio of 2:1 Manx: normal. The homozygous dominant condition (M^LM^L) is **lethal** and embryos with this genotype are resorbed and never appear.

3. Some dominant lethal alleles, including Huntington's, do not take effect until after the onset of adulthood, i.e. after those with the genotype have reached reproductive maturity. There is a chance to pass the allele onto children before it takes effect.

Problems in Mendelian Genetics (page 177)

1. **Note:** Persian and Siamese parents are pedigrees (truebreeding) and homozygous for the genes involved.
(a) Persian: UUss, Siamese: uuSS, Himalayan: uuss
(b) F_1: Genotype: all heterozygotes UuSs.
(c) F_1: Phenotype: all uniform color, short haired.
(d) F_2 generation: UuSs **X** UuSs

	US	**Us**	**uS**	**us**
US	UUSS	UUSs	UuSS	UuSs
Us	UUSs	UUss	UuSs	Uuss
uS	UuSS	UuSs	uuSS	uuSs
us	UuSs	Uuss	uuSs	uuss

(e) 1:15 or 1/16 uuss: Himalayan
(f) Yes (only one type of allele combination is possible)
(g) 3:13 or 3/16 (2 uuSs, 1 uuSS)
(h) All of the following have different genotypes but produce a uniform color-short hair cat: UUSS, UuSS, UuSs, UUSs, because they all have at least one dominant allele for each gene. Similarly uuSs and uuSS both produce a color pointed short hair cat, and UUss and Uuss both produce a uniform colored, long hair cat.
(i) Test cross with a Himalayan (i.e. double homozygous recessive: uuss). If any heterozygous cats are presented for mating then some of their offspring could be expected to be long-haired.

2. (a) **bbSS** (brown/spotted) X **BBss** (solid/black) (which parent was male and which female is unknown. Parents must be homozygous since all the offspring are of one type: BbSs: black spotted)
(b) F_2 generation: BbSs **X** BbSs

	BS	**Bs**	**bS**	**bs**
BS	BBSS	BBSs	BbSS	BbSs
Bs	BBSs	BBss	BbSs	Bbss
bS	BbSS	BbSs	bbSS	bbSs
bs	BbSs	Bbss	bbSs	bbss

(c) Spotted/black 9/16

Spotted/brown	3/16
Solid/black	3/16
Solid/brown	1/16
Ratio:	9:3:3:1 (described as above)

(d) Dihybrid cross (no linkage)

3. (a) F_1: Genotype: all heterozygotes RrBb.
(b) F_1: Phenotype: all rough black coats.
(c) F_2 generation: RrBb **X** RrBb

	RB	**Rb**	**rB**	**rb**
RB	RRBB	RRBb	RrBB	RrBb
Rb	RRBb	RRbb	RrBb	Rrbb
rB	RrBB	RrBb	rrBB	rrBb
rb	RrBb	Rrbb	rrBb	rrbb

(d)

Rough/black	9/16
Rough/white	3/16
Smooth/black	3/16
Smooth/white	1/16
Ratio:	9:3:3:1 (described as above)

(e) F_2 generation: RrBb **X** RRBB

	RB	**RB**	**RB**	**RB**
RB	RRBB	RRBB	RRBB	RRBB
Rb	RRBb	RRBb	RRBb	RRBb
rB	RrBB	RrBB	RrBB	RrBB
rb	RrBb	RrBb	RrBb	RrBb

(f) F_2 Phenotype: all rough black coats.
(g) F_2 generation: RrBb **X** rrbb

	rb	**rb**	**rb**	**rb**
RB	RRBb	RRBb	RRBb	RRBb
Rb	Rrbb	Rrbb	Rrbb	Rrbb
rB	rrBb	rrBb	rrBb	rrBb
rb	rrbb	rrbb	rrbb	rrbb

(h) Note that this is also a back cross, since the cross is back to the parental phenotype.

Rough/black	4/16 (RrBb)
Rough/white	4/16 (Rrbb)
Smooth/black	4/16 (rrBb)
Smooth/white	4/16 (rrbb)
Ratio:	1:1:1:1 (described as above)

(i) The parent's genotype: RrBb **X** Rrbb

4. The homozygous condition (C^LC^L) is lethal. Normal birds are CC. The alleles show incomplete dominance, which is why the heterozygous condition (CC^L) produces creeper birds with a particular phenotype (short wings/legs). A cross between two creepers produces two creepers to every one normal bird (not three creepers as would be expected if the allele was not lethal in its homozygous condition).

5. Probability of black offspring: (2/3 x 1/4=) 1/6 or 0.16

Working: The parents genotypes are Bb X Bb, and 1/3 of the white offspring (BB) crossed with **Bb** will result in no black lambs while 2/3 of the white offspring (Bb) crossed with Bb will result in 1/4 black lambs.

6. (a) Genotypes: 1/2 M^RM^R, 1/2 M^RM
Phenotypes: All restricted mallard pattern
(b) Genotypes: 1/2 M^RM^R, 1/2 M^Rm
Phenotypes: All restricted mallard pattern
(c) Genotypes: 1/4 M^RM^R, 1/4 M^RM, 1/4 M^Rm, 1/4 Mm
Phenotypes: 3/4 restricted mallard pattern, 1/4 mallard pattern
Ratio: 3 restricted mallard : 1 mallard pattern
(d) Genotypes: 1/4 M^RM, 1/4 M^Rm, 1/4 Mm, 1/4 mm
Phenotypes: 1/2 restricted mallard pattern, 1/4 Mallard pattern, 1/4 dusky mallard pattern
Ratio: 2 restricted mallard : 1 mallard :1 dusky mallard pattern
(e) Genotypes: 1/2 Mm, 1/2 mm
Phenotypes: 1/2 mallard pattern, 1/2 dusky mallard pattern
Ratio: 1 mallard : 1 dusky mallard pattern

7. 1/2 Ww and 1/2 ww
Ratio: 1 wire-haired : 1 smooth haired
Working: The parental genotypes are Ww X Ww. The test cross of the F1 is to a smooth haired dog (**ww**). 1/4 of the F1 will be wire-haired dogs (WW). When crossed with ww the result will be all wire-haired dogs (Ww).
1/2 of the F1 will be wire-haired dogs (Ww). When crossed with ww, the result will be 1/2 wire-haired and 1/2 smooth-haired dogs.
1/4 of the F1 will be smooth-haired dogs (ww) . When crossed with ww, all offspring will smooth-haired (ww).

Human Genotypes (page 179)

This will vary with each individual's collection of genes, e.g.

Thumb	Hh	Hyperextension
Ear lobes	FF	Free
Chin cleft	DD	Dimpled
Middle digit hair	Mm	Hair
Handedness	RR	Right
Hand clasp	cc	Right thumb on top

Sex Determination (page 181)

1. Presence of the Y chromosome (XX female, XY male).
2. The males have differing sex chromosomes (X and Y).

Genomic Imprinting (page 182)

1. (a) **Genomic imprinting** (or parental imprinting) is part of epigenetics, the study of heritable changes in gene function that occur without involving changes in the DNA sequence. It occurs during gametogenesis in which the expression of a small subset of genes depends on whether the genes are inherited from the mother or father (the parent-of-origin effect).
(b) Imprinting is achieved by (any one of):
– DNA methylation of the allele contributed by one parent, making an offspring effectively homozygous for the other parent's allele.
– Inheritance of two homologous copies of a gene from one parent (called uniparental disomy).

2. Imprinting is significant to the inheritance of some genes as imprinting will affect their phenotypic expression. For example, the same mutation, a specific deletion on chromosome 15, produces two different human genetic disorders depending on whether the mutation is inherited from the mother or the father.

Sex Linkage (page 183)

1.

Parent genotype:	X_oX_o	X_OY
Gametes:	X_o, X_o	X_O, Y
Kitten genotypes:	X_OX_o, X_oY	X_OX_o, X_oY
	Genotypes	**Phenotypes**
Male kittens:	X_oY	Black
Female kittens:	X_OX_o	Tortoiseshell

2.

Parent genotype:	X_OX_o	X_OY
Gametes:	X_o, X_O	X_O, Y
Kitten genotypes:	X_OX_O, X_OX_o	X_oY, X_OY
Phenotypes:	Orange female, Tortoise female	Black male, Orange male

(a) Father's genotype: X_OY
(b) Father's phenotype: Orange

3.

Parent genotype:	X_oX_o	X_OY
Gametes:	X_o, X_o	X_O, Y
Kitten genotypes:	X_OX_o, X_oY	X_OX_o, X_oY
Phenotypes:	Tortoise female, Black male	Tortoise female, Black male

(a) Father's genotype: X_OY
(b) Father's phenotype: Orange
(c) Yes, the same male cat could have fathered both litters.

4.

Parent:	Normal wife	Affected husband
Parent genotype:	XX	X_RY
Gametes:	X, X	X_R, Y
Children's genotypes:	X_RX, XY	X_RX, XY
Phenotypes:	Affected girl, Normal boy	Affected girl, Normal boy

(a) Probability of having affected children = 50% or 0.5
(b) Probability of having an affected girl = 50% or 0.5
However, all girls born will be affected = 100%
(c) Probability of having an affected boy = 0% or none

5.

Parent:	Affected wife	Normal husband
Parent genotype:	X_RX	XY
Gametes:	X_R, X	X, Y
Children's genotype:	X_RX, X_RY	XX, XY
Phenotypes:	Affected girl, Affected boy	Normal girl, Normal boy

Note: Because the wife had a normal father, she must be heterozygous since her father was able to donate only an X-chromosome with the normal condition.
(a) Probability of having affected children = 50% or 0.5
(b) Probability of having an affected girl = 25% or 0.25
However, half of all girls born may be affected.
(c) Probability of having an affected boy = 25% or 0.25
However, half of all boys born may be affected.

Background information for question 6: Sex linkage refers to the location of genes on one or other of the sex chromosomes (usually the X, but a few are carried on the Y). Such genes produce an inheritance pattern which is different from that shown by autosomes:
- Reciprocal crosses produce different results (unlike autosomal genes that produce the same results).
- Males carry only one allele of each gene.
- Dominance operates in females only.
- A 'cross-cross' inheritance pattern is produced: father to daughter to grandson, etc.

6. Sex linkage is involved in a number of genetic disorders (below). X-linked disorders are commonly seen only in males (the heterogametic sex), because they have only one locus for the gene and must express the trait. If the sex linked trait is due to a recessive allele, females will express the phenotype only when homozygous recessive. It is possible for females to inherit a double dose of the recessive allele (e.g. a color blind daughter can be born to a color blind father and mother who is a carrier), but this is much less likely than in males. X-linked genes include those that control:
- Blood clotting: A recessive allele for this gene causes hemophilia. It affects about 0.01% of males but is almost unheard of in females.
- Normal color vision: A recessive allele causes red-green color blindness affecting 8% of males but only 0.7% of females.
- Antidiuretic hormone production: A version of this gene causes some forms of diabetes insipidus.
- Muscle development: A rare recessive allele causes Duchene muscular dystrophy.

Inheritance Patterns (page 185)

1. **Autosomal recessive:**
(a) Punnett square:
Male parent phenotype:
Normal, carrier
Female parent phenotype:
Normal, carrier

	P	p
P	PP	Pp
p	Pp	pp

(b) Phenotype ratio:
Normal 3 Albino 1

2. **Autosomal dominant:**
(a) Punnett square:
Male parent phenotype:
Woolly hair
Female parent phenotype:
Woolly hair

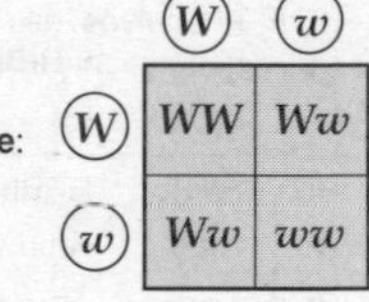

	W	w
W	WW	Ww
w	Ww	ww

(b) Phenotype ratio:
Normal 1 Woolly 3

3. **Sex linked recessive:**
(a) Punnett square:
Male parent phenotype:
Normal
Female parent phenotype:
Normal, carrier

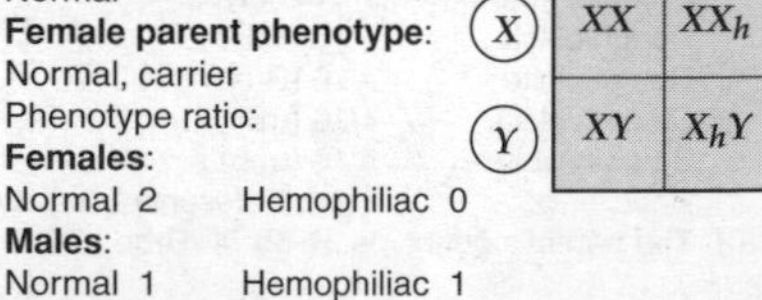

	X	X_h
X	XX	XX_h
Y	XY	X_hY

(b) Phenotype ratio:
Females:
Normal 2 Hemophiliac 0
Males:
Normal 1 Hemophiliac 1

4. **Sex linked dominant:**
(a) Punnett square:
Male parent phenotype:
Affected (with rickets)
Female parent phenotype:
Affected (with rickets)

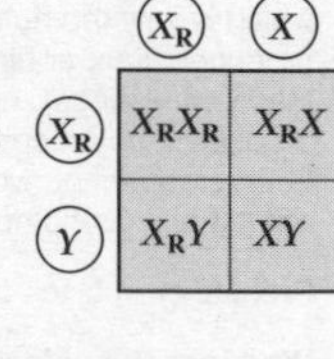

	X_R	X
X_R	X_RX_R	X_RX
Y	X_RY	XY

(b) Phenotype ratio:
Females:
Normal 0 Rickets 2
Males:
Normal 1 Rickets 1

Pedigree Analysis (page 186)

1. **Pedigree chart of your family**: The chart should be drawn up using the appropriate symbols.

2. **Autosomal recessive traits**
(a) Genotypes of individuals on the chart:

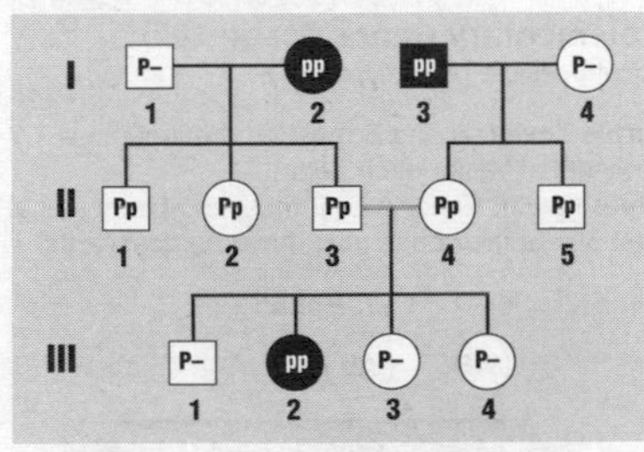

(b) The genotypes of parents II-3 and II-4 have to be carriers (heterozygous) because they produced an affected offspring (homozygous recessive). Alternatively, we know that they must be carriers because each had an affected parent (I-2 and I-3).

3. **Sex-linked recessive traits**
(a) Genotypes of individuals on the chart:

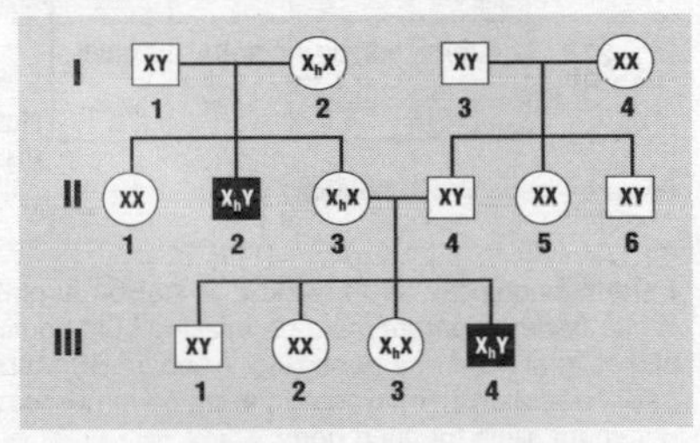

(b) Males have only one X chromosome. If that single chromosome carries the affected allele, then it will be expressed. Males cannot be heterozygous, since they can only carry one copy of the gene.

4. **Autosomal dominant traits**
(a) Genotypes of individuals on the chart:

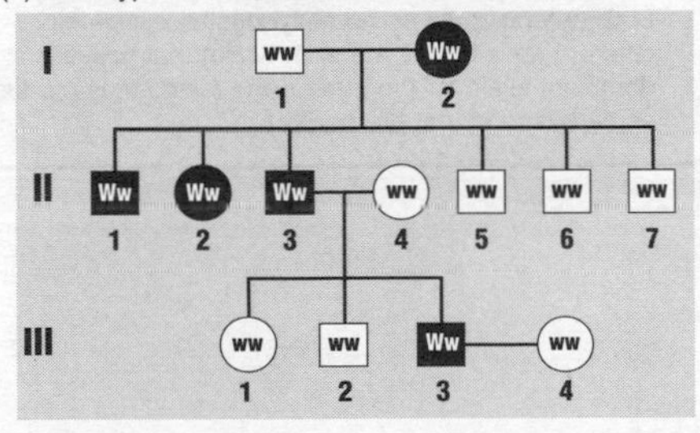

Teacher's note: Parent I-2 has to be heterozygous (Ww) since some of the offspring are normal (ww). This could not occur if I-2 was homozygous dominant (WW).

(b) Each affected individual has an affected parent.

5. **Sex-linked dominant traits**
(a) Genotypes of individuals on the chart:

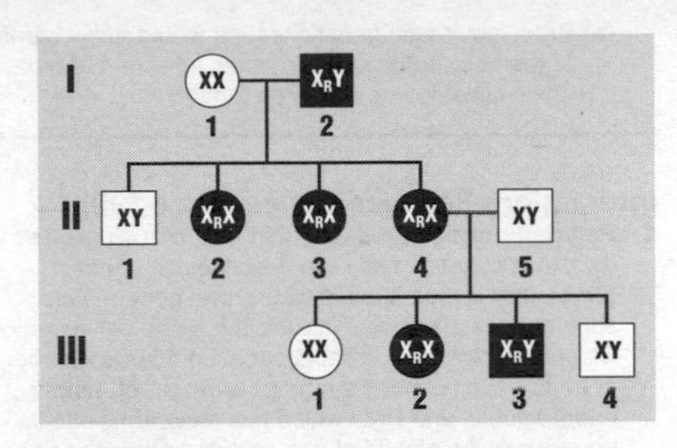

(b) Females have two X chromosomes, and therefore have a greater probability that one of them will carry the affected gene.

6. (a) Dominant trait. Each individual has an affected parent.
(b) Not sex-linked. Some of the daughters of the affected parent (I-3) are not affected. They would be if the gene were located on the X chromosome.

7. (a) Explanatory background: I-1 is a normal male, while the mother (I-2) must be a carrier (heterozygous) because she gave birth to both a normal and a haemophiliac son. There is a 50% probability that II-2 is heterozygous (she could have received either of her mother's X chromosomes). If a carrier woman has children by a normal man then 25% of their children can be expected to be haemophiliacs. Because there is uncertainty as to whether the woman is a carrier or not, the total probability of II-2 producing haemophiliac children is: Probability of being a carrier X probability of producing haemophiliac children if she is a carrier i.e. **1/2 x 1/4 = 1/8 (12.5%).**
(b) 1/4 (25%). Because her first child was a haemophiliac, she must be a carrier.
(c) 3/4 (75%). II-4 has a 50% chance of being a carrier. If she was a carrier and has children with a haemophiliac man, 1/2 of their children (boys and girls) are expected to be haemophiliac. The combined chance of II-4 being a carrier and producing a haemophiliac child is 1/2 X 1/2 = 1/4. Therefore the probability that the child will be normal is the complementary fraction (3/4).
(d) It is impossible to determine the phenotype of the father of I-1 from the information given, because the father could be either normal or a haemophiliac and still produce a daughter (I-1) which is a carrier (heterozygous normal).

8. (a) 1/2 (b) 0 (c) 0
(d) 3/4 (e) 1/2

Genetic Counseling (page 189)

1. **Carrier screening** would allow the parents to ascertain the risk of conceiving a child with the family genetic disorder. This information could then be used to explore alternative methods of conceiving a child that is at lower risk of inheriting the disorder.

2. (a) Huntington disease persists because those with the mutation are able to reproduce and pass on the disorder before its effects are expressed.

(b) Presymptomatic genetic testing would allow carriers of the Huntington gene to make informed choices when considering starting a family.

Interactions Between Genes (page 190)

1. **Polygeny** refers to the determination of a single trait (e.g. skin color) by two or more genes. In contrast, **pleiotropy** is the situation where one gene affects several traits. Examples include the sickle cell mutation, which has pleiotropic effects because a large number of traits are influenced by the possession of mutant haemoglobin, and PKU which has pleiotropic effects because of the effects of accumulated phenylalanine and its derivative phenylpyruvate.
2. **Epistasis** describes the situation where one gene masks or alters the expression of other genes. In contrast, a pleiotropic gene has multiple phenotypic effects but does not alter the expression of other genes. Male pattern baldness is an example of epistasis: if expressed, baldness overrides the expression of genes determining hairline (e.g. widow's peak).
3. If a genotype is homozygous recessive for albinism, the expression of that gene (i.e no melanin produced) is epistatic to all other genes for color.
4. PKU results in an inability to convert phenylalanine to tyrosine, and phenylalanine concentrations increase to toxic levels. This causes damage at several locations in the body and leads to the pleiotropic effects typical of the disease: mental retardation, reduced hair and skin pigmentation, and dental disorders.

Collaboration (page 191)

1. No. of possible phenotypes: 4
2. Pea comb: rrP_
   Rose comb: R_pp
   Walnut comb: R_P_
   Single comb: rrpp
3. See the Punnett square below.

Sperm (columns) / Eggs (rows)

Eggs \ Sperm	RP	Rp	rP	rp
RP	RRPP Walnut	RRPp Walnut	RrPP Walnut	RrPp Walnut
Rp	RRPp Walnut	RRpp Rose	RrPp Walnut	Rrpp Rose
rP	RrPP Walnut	RrPp Walnut	rrPP Pea	rrPp Pea
rp	RrPp Walnut	Rrpp Rose	rrPp Pea	rrpp Single

Ratio: Walnut : Pea : Rose : Single
9 : 3 : 3 : 1

4. Parent genotypes:
   (a) Rrpp (rose) X RrPp (walnut)
   (b) rrpp (single) X RrPp (walnut)
   (c) RRpp (rose) X rrPp (pea)

Complementary Genes (page 192)

1. No. of possible phenotypes: 2
2. **Purple flower**: A and B must be present (i.e. a dominant allele for each gene).
   White flower: Either A or B must be absent (i.e. at least one of the genes must have no dominants).
3.

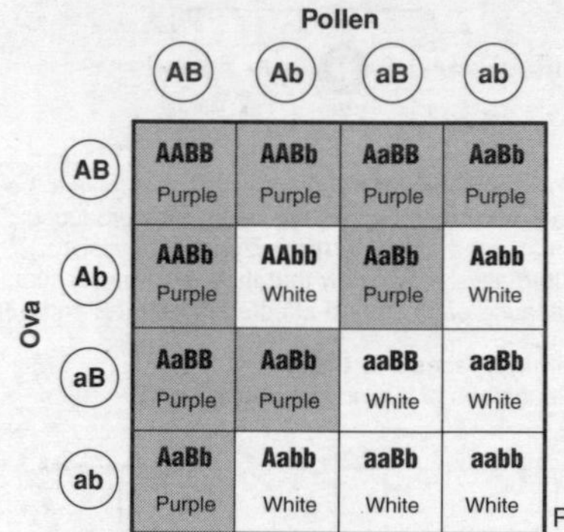

Pollen (columns) / Ova (rows)

Ova \ Pollen	AB	Ab	aB	ab
AB	AABB Purple	AABb Purple	AaBB Purple	AaBb Purple
Ab	AABb Purple	AAbb White	AaBb Purple	Aabb White
aB	AaBB Purple	AaBb Purple	aaBB White	aaBb White
ab	AaBb Purple	Aabb White	aaBb White	aabb White

Ratio: Purple flowers : White flowers
9 : 7

4. Parent genotypes: aaBb (white) X AaBb (purple)
   White flowered parent must be missing both dominant alleles in at least one gene (e.g. AAbb, aaBB, Aabb, aaBb, aabb). Purple flowered parent must possess a dominant allele for each gene.
5. Parent genotypes: AaBb (purple) X AaBb (purple)
   Both parents had to possess a copy of the dominant allele for each gene in order to be purple. They also must have a recessive allele for each gene in order to produce some offspring that are white.
6. Parent genotypes: aaBb (white) X Aabb (white).
   Both parents must be homozygous recessive for different genes (they are both white), but possess a dominant allele for the other gene (since they produce some offspring that are purple).

Polygenes (page 193)

1. Punnett square

Gametes	ABC	ABc	AbC	Abc	aBC	aBc	abC	abc
ABC	AABBCC	AABBCc	AABbCC	AABbCc	AaBBCC	AaBBCc	AaBbCC	AaBbCc
ABc	AABBCc	AABBcc	AABbCc	AABbcc	AaBBCc	AaBBcc	AaBbCc	AaBbcc
AbC	AABbCC	AABbCc	AAbbCC	AAbbCc	AaBbCC	AaBbCc	AabbCC	AabbCc
Abc	AABbCc	AABbcc	AAbbCc	AAbbcc	AaBbCc	AaBbcc	AabbCc	Aabbcc
aBC	AaBBCC	AaBBCc	AaBbCC	AaBbCc	aaBBCC	aaBBCc	aaBbCC	aaBbCc
aBc	AaBBCc	AaBBcc	AaBbCc	AaBbcc	aaBBCc	aaBBcc	aaBbCc	aaBbcc
abC	AaBbCC	AaBbCc	AabbCC	AabbCc	aaBbCC	aaBbCc	aabbCC	aabbCc
abc	AaBbCc	AaBbcc	AabbCc	Aabbcc	aaBbCc	aaBbcc	aabbCc	aabbcc

Darker skin   Same   Lighter skin

(a) 20   (b) 27

2. Environmental influences will alter the color of a person skin (such as tanning) to different extents.

3. Traits with **continuous variation** show a normal distribution curve when sampled and a graded variation in phenotype in the population. Such phenotypes are usually determined by a large number of genes and/ or environmental influence. Examples include height, weight, hand span, foot size. In contrast, traits with **discontinuous variation** fall into one of a limited number of phenotypic variants and do not show a normal distribution curve when sampled. Differences in the phenotypes of individuals in a population are marked and do not grade into each other. Such phenotypes are usually controlled by a few different alleles at a few genes. Examples include ear lobe shape and tongue roll.

4. Student's own plot. Shape of the distribution is dependent on the data collected. The plot should show a **statistically normal distribution** if sample is representative of the population and large enough.
   (a) Calculations based on the student's own data.
   (b) Continuous distribution, normal distribution, or bell shaped curve are all acceptable answers if the data conform to this pattern.
   (c) Polygenic inheritance: Several (two or more) genes are involved in determining the phenotypic trait. Environment may also have an influence, especially if traits such as weight are chosen.
   (d) A large enough sample size (30+) provides sufficient data to indicate the distribution. The larger the sample size, the more closely one would expect the data plot to approximate the normal curve (assuming the sample was drawn from a population with a normal distribution for that attribute).

Epistasis (page 195)

1. No. of phenotypes (… for this **type** of epistasis): 3

2. Black: B_C_ (a dominant allele for each gene)
   Brown: A dominant allele for gene C only (e.g. Ccbb)
   Albino: No dominant allele for gene C (e.g. ccBB, ccb

3.

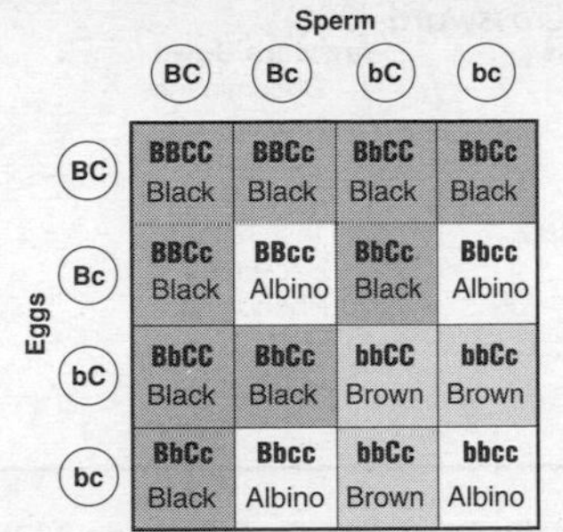

Eggs \ Sperm	BC	Bc	bC	bc
BC	BBCC Black	BBCc Black	BbCC Black	BbCc Black
Bc	BBCc Black	BBcc Albino	BbCc Black	Bbcc Albino
bC	BbCC Black	BbCc Black	bbCC Brown	bbCc Brown
bc	BbCc Black	Bbcc Albino	bbCc Brown	bbcc Albino

Ratio: Black: brown: albino
9: 3: 4

4. Homozygous albino (bbcc) x homozygous black (BBCC):
   Offspring genotype: 100% BbCc
   Offspring phenotype: 100% black

5. Homozygous brown (bbCC) x homozygous black (BBCC):
   Offspring genotype: 100% BbCC
   Offspring phenotype: 100% black

6. Four phenotypes: (1) black, (2) chocolate, (3) yellow with a black nose, eyebrows, and lips, and (4) yellow with a brown nose, eyebrows, and lips.

7. Black: dominant E and dominant B alleles
   Brown (chocolate): dominant E, recessive bb
   Yellow with brown nose: homozygous recessive for both genes (ee and bb)
   Yellow with black nose: recessive ee and dominant B.

8.

Eggs \ Sperm	EB	Eb	eB	eb
EB	EEBB Black	EEBb Black	EeBB Black	EeBb Black
Eb	EEBb Black	EEbb Choc	EeBb Black	Eebb Choc
eB	EeBB Black	EeBb Black	eeBB Yellow w/ black	eeBb Yellow w/ black
eb	EeBb Black	Eebb Choc	eeBb Yellow w/ black	eebb Yellow w/ brown

9 Black, 3 chocolate, 3 yellow with black, 1 yellow with brown

9. Chocolate (EEbb) X Black (EEBB)
   Offspring EEBb. All black Labradors

10. Yellow with brown nose (eebb) X black (EEBB)
    Offspring EeBb All black Labradors.

What Genotype Has That Cat? (page 199)

Model answers cannot be provided for this exercise as there are so many different possible cat phenotypes. After the *Phenotype Record Sheet* has been completed, the extension exercise, where the genotype of each cat is identified should be completed on a separate sheet.

KEY TERMS: Crossword (200)

Answers Across
2. Multiple
6. Inheritance
8. Autosome
10. Pleiotropy
12. Complementary
13. Pedigree
14. Incomplete
16. Locus
17. Polygenic
18. Genotype

Answers Down
1. Codominance
3. Linkage
4. Recessive
5. Epigenetics
7. Testcross
9. Backcross
10. Phenotype
11. Trait
15. Cline

Does DNA Really Carry The Code? (page 202)

1. Griffith first injected disease causing bacteria into healthy mice to confirm the action of the bacteria. The bacteria were then heated to kill them and injected into healthy mice. These mice did not develop pneumonia confirming it was the living bacteria that was causing the disease.

2. Sulfur is found in proteins but not in DNA (or to a much lesser extent) while phosphorus is found in DNA but not proteins. If the sulfur was found in the infected bacteria then the proteins carried the genetic information. If phosphorus was found then it was the DNA that carried the information.

3. This shows that the result is not peculiar to, or the result of, the experiment itself. Rather it is a property of the system being studied.

DNA Molecules (page 203)

1. (a) 95 times more base pairs
   (b) 630 times more base pairs

2. < 2% encodes proteins or structural RNA.

3. (a) and (b) in any order:
   (a) Much of the once considered 'junk DNA' has now been found to give rise to functional RNA molecules (many with regulatory functions).
   (b) Complex organisms contain much more of this non-protein-coding DNA which suggests that these sequences contain RNA-only 'hidden' genes that have been conserved through evolution and have a definite role in the development of the organism.

The Genetic Code (page 204)

1. This exercise demonstrates the need for a 3-nucleotide sequence for each codon and the 'degeneracy' in the genetic code.

Amino acid	Codons						No
Alanine	GCU	GCC	GCA	GCG			4
Arginine	CGU	CGC	CGA	CGG	AGA	AGG	6
Asparagine	AAU	AAC					2
Aspartic Acid	GAU	GAC					2
Cysteine	UGU	UGC					2
Glutamine	CAA	CAG					2
Glutamic Acid	GAA	GAG					2
Glycine	GGU	GGC	GGA	GGG			4
Histidine	CAU	CAC					2
Isoleucine	AUU	AUC	AUA				3
Leucine	UAA	UUG	CUU	CUC	CUA	CUG	6
Lysine	AAA	AAG					2
Methionine	AUG						1
Phenylalanine	UUU	UUC					2
Proline	CCU	CCC	CCA	CCG			4
Serine	UCU	UCC	UCA	UCG	AGU	AGC	6
Threonine	ACU	ACC	ACA	ACG			4
Tryptophan	UGG						1
Tyrosine	UAU	UAC					2
Valine	GUU	GUC	GUA	GUG			4

2. (a) 16 amino acids
   (b) Two-base codons (e.g. AT, GG, CG, TC, CA) do not give enough combinations with the 4-base alphabet (A, T, G and C) to code for the 20 amino acids.

3. Many of the codons for a single amino acid vary in the last base only. This would reduce the effect of point mutations; only some changes would create new and potentially harmful amino acid sequences. **Note**: Only 61 codons are displayed above. The remaining 3 are **terminato**r codons (labeled 'STOP' codons in the table in the workbook). These are considered the 'punctuation' or controlling codons that mark the end of a gene sequence. The amino acid **methionine** (AUG) is regarded as the 'start' (initiator) codon.

Creating a DNA Model (page 205)

3. Labels as follows:

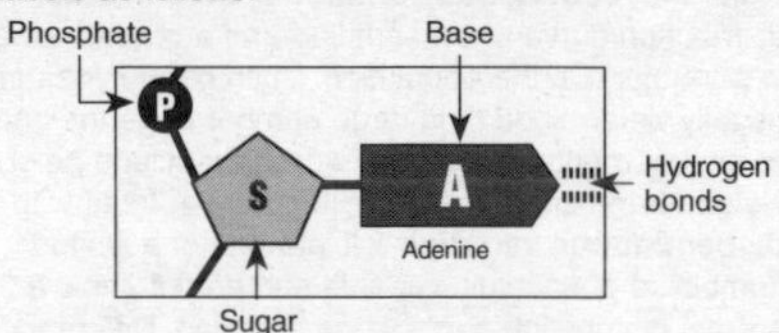

4. & 5. See the next column.

6. Factors preventing a mismatch of nucleotides:
   (a) The number of hydrogen bond attraction points.
   (b) The size (length) of the base (thymine and cytosine are short, adenine and guanine are long).

   Examples:
   - Cytosine will not match cytosine because the bases are too far apart.
   - Guanine will not match guanine because they are too long to fit side-by-side.
   - Thymine will not match guanine because there is a mis-match in the number and orientation of H bonds.

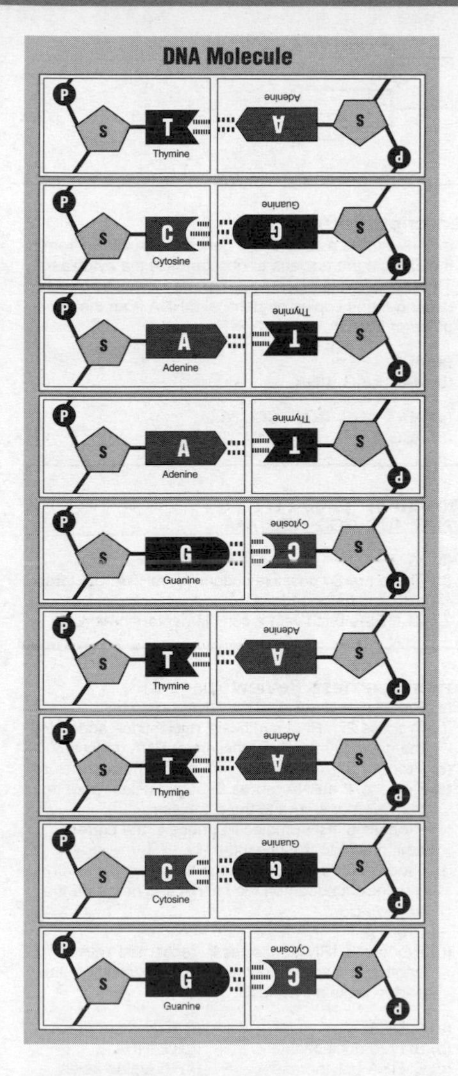

DNA Replication (page 209)

1. DNA replication prepares a chromosome for cell division by producing two chromatids which are (or should be) identical copies of the genetic information for the chromosome.

2. (a) Step 1: Enzymes unwind DNA molecule to expose the two original strands.
   (b) Step 2: DNA polymerase enzyme uses the two original strands as templates to make complementary strands.
   (c) Step 3: The two resulting double-helix molecules coil up to form two chromatids in the chromosome.

3. (a) Helicase: Unwinds the 'parental' strands.
   (b) DNA polymerase I: Hydrolyzes the RNA primer and replaces it with DNA.
   (c) DNA polymerase III: Elongates the leading strand. It synthesizes the new Okazaki fragment until it encounters the primer on the previous fragment.
   (d) Ligase: Joins Okazaki fragments into a continuous length of DNA.

4. 16 minutes 40 seconds
   4 million nucleotides replicated at the rate of 4000 per second: 4 000 000 ÷ 4000 = 1000 s
   Convert to minutes = 1000 ÷ 60 = 16.67 minutes
   (Note that, under ideal conditions, most of a bacteria's cell cycle is spent in cell division).

A Review of DNA Replication (page 211)

1. (a) G pairs with C
   A pairs with T
   (b) A: Parent DNA
   B: Swivel point or replication fork
   C: DNA Polymerase III
   D: Parent strand of DNA in the new DNA molecule
   E: Daughter strand of DNA in the new DNA molecule
   F: Free nucleotide

3. Steps 1-4 as follows:
   1: DNA strands are joined by base pairing.
   2: Unwinding of parent DNA double helix.
   3: Unzipping of DNA parent.
   4: Free nucleotides occupy spaces along exposed bases.

Note : In the next activities, including that on transcription, the **accepted convention** amongst molecular geneticists has been used, i.e. the **coding** (sense or non-transcribed) strand contains the **same base sequence as the mRNA** that is transcribed from the **template** (antisense) **strand.** This terminology arises from the fact of where the gene (to be transcribed) is located (the "coding" strand). This may oppose what appears in some texts, where there is much confusion in the use of these terms. In fact, with the exception of template strand, the modern view (and the view in the more authoritative, current texts) is to avoid the use of these terms, as they imply that one strand alone always carries the genes.

The Simplest Case: Genes to Proteins (page 212)

1. This exercise shows the way in which DNA codes for proteins. Nucleotide has no direct protein equivalent.
   (a) Triplet codes for amino acid.
   (b) Gene codes for polypeptide chain (may be a polypeptide, protein, or RNA product).
   (c) Transcription unit codes for functional protein.

2. (a) **Nucleotides** are made up of: phosphate, sugar, and one of four bases (adenine, guanine, cytosine, and thymine or uracil).
   (b) **Triplet** is made up of three consecutive nucleotide bases that are read together as a code.
   (c) **Gene** comprises a sequence of triplets, starting with a start code and ending with a termination code.
   (d) **Transcription unit** is made up of two or more genes that together code for a functional protein.

3. Steps in making a functional protein:
   - The template strand is made from the DNA coding

strand and is transcribed into mRNA.
- The code on the mRNA is translated into a sequence of amino acids, which are linked with peptide bonds to form a polypeptide chain (this may be a functional protein in its own right).
- The proteins coded by two or more genes come together to form the final functional protein.

Gene Expression (page 213)

1. In **prokaryotic** gene expression, RNA is translated into protein almost as fast as it is transcribed; there is no nucleus therefore no separation of the transcription and translation processes. The presence of introns in a prokaryotic genome would interfere with protein function because there would be no time between transcription and translation to splice them out. In contrast, **eukaryotic** gene expression involves production of a primary RNA transcript from which the introns are removed. There is sufficient time for this to take place because transcription and translation occur inside and outside the nucleus respectively. **Note**: Evidence in support of this: Prokaryote DNA consists almost entirely of protein-coding genes and their regulatory sequences with very little non-protein coding DNA. Eukaryotic DNA comprises large amounts of non-protein coding sequences, much of which we now know codes for functional (regulatory) RNAs.
2. In the new view of eukaryotic gene expression, the so-called "junk DNA", traditionally regarded as non-functional, is considered to play a necessary role in the eukaryote cell. In fact, there is a good correlation between the amount of non-protein coding DNA a species has and its complexity. Although this DNA does not code for proteins, its does code for RNA molecules with regulatory functions (some of these involving regulation of the genome itself). What's more, in the new view of gene expression, not all of the exonic RNA is translated into protein; some contributes to regulatory microRNAs. In the old view, all exonic RNA was thought to code for proteins.
3. The one gene-one protein model is still appropriate for prokaryotes because, in the absence of introns, RNA transcripts are translated directly into proteins. Because of the small size of prokaryotic genomes and their one site of protein synthesis, there is very little non-protein coding DNA present.

Analyzing a DNA Sample (page 215)

1. Use the *mRNA* table on the page: **The Genetic Code** in the workbook to determine the amino acid sequence.

Synthesized DNA	CGT AAG TAC TTG ATC AGA GCT CTT CGA AAA TCG
DNA sample:	GCA TTC ATG AAC TAG TCT CGA GAA GCT TTT AGC
mRNA:	CGU AAG UAC UUG AUC AGA GCU CUU CGA AAA UCG
Amino acids:	Arg Lys Tyr Leu Iso Arg Ala Leu Arg Lys Ser

2. (a) ATG ATC GGC GCT AAA TGT TAA
   (b) ATG CGG AAT TTC CCG GCT TAG
   (c) DNA replication

3. 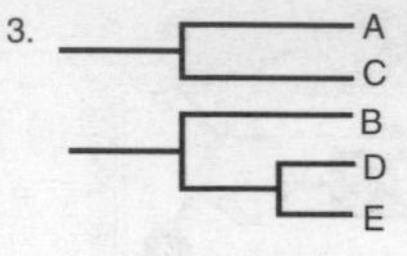

Transcription (page 216)

1. mRNA carries a copy of the genetic instructions from the DNA in the nucleus to ribosomes in the cytoplasm. The rate of protein synthesis can be increased by making many copies of identical mRNA from the same piece of DNA.
2. (a) AUG
   (b) UAA, UAG, UGA
3. (a) AUG AUC GGC GCU AAA
   (b) AUG UUC GGA UAU UUU

Translation (page 217)

1. AUG AUC GGC GCU AAA
2. (a) 61
   (b) There are 64 possible codons for mRNA, but three are terminator codons. 61 codons for mRNA require 61 tRNAs each with a complementary codon.

Protein Synthesis Review (page 218)

1. (a) Process 1: Unwinding the DNA molecule.
   (b) Process 2: mRNA synthesis, nucleotides added to the growing strand of messenger RNA molecule.
   (c) Process 3: DNA rewinds into double helix structure.
   (d) Process 4: mRNA moves through nuclear pore in nuclear membrane to the cytoplasm.
   (e) Process 5: tRNA molecule brings in the correct amino acid to the ribosome.
   (f) Process 6: Anti-codon on the tRNA matches with the correct codon on the mRNA and drops off the amino acid.
   (g) Process 7: tRNA leaves the ribosome.
   (h) Process 8: tRNA molecule is 'recharged' with another amino acid of the same type, ready to take part in protein synthesis.
2. (a) A: DNA
   (b) B: Free nucleotides
   (c) C: RNA polymerase
   (d) D: mRNA
   (e) E: Nuclear membrane
   (f) F: Nuclear pore
   (g) G: tRNA
   (h) H: Amino acids
   (i) I: Ribosome
   (j) J: Polypeptide chain
3. Factors determining whether or not a protein is produced (in any order):
   (a) Whether or not the protein is required by the cell (regulated by control of gene expression).
   (b) Whether or not there is an adequate pool of the amino acids and tRNAs required for the particular protein in question.

Control of Metabolic Pathways (page 219)

1. A **metabolic pathway** is simply a series of related chemical reactions that converts one compound to another in a sequence of steps. Each step in a metabolic pathway is controlled by enzymes; the

end product of one enzyme-controlled step provides the substrate for the next step. In the metabolism of the amino acid phenylalanine, an enzyme controls the conversion of phenylalanine to tyrosine. Enzyme controlled steps lead to thyroxine (a hormone), to melanin, or to any of a series of intermediate breakdown products. Failure of enzymes in any of these many steps leads to clearly defined metabolic disorders of varying severity.

2. Any three of: thyroxine, melanin, carbon dioxide, water
3. (a) Tyrosinase
   (b) Phenylalanine hydroxylase
   (c) Hydroxyphenylpyruvic acid oxidase
   (d) Homogentisic acid oxidase
4. People with PKU have low levels of tyrosine, the raw material for making melanin (the pigment that gives dark color to the skin and hair). Tyrosine is normally created from phenylalanine by an enzyme, which is faulty in this case.
5. Disorders in tyrosine metabolism can lead to cretinism or albinism. Cretinism is the result of errors in the metabolic pathway to thyroxine, an important growth hormone for the development of body organs. The lack of thyroxine results in the symptoms of cretinism: small body size (dwarfism), mental retardation, and undeveloped sexual organs. Albinism is the result of malfunctions of the enzyme tyrosinase, which converts tyrosine to the pigment melanin. Lack of melanin results in the symptoms of albinism: a total lack of pigmentation in the skin, eyes, and hair.
6. (a) Lack of melanin.
   (b) Excess phenylpyruvic acid.
   (c) Lack of thyroxine.
   (d) Excess hydroxyphenylpyruvic acid.
   (e) Excess homogentisic acid.
7. You would take a blood sample and test for excessive amounts of phenylpyruvic acid (in fact, this is a routine blood test performed five days after birth).
8. (a) Chemicals present in excess: Precursor
   (b) Chemicals absent: Intermediate, end product

Gene Control in Prokaryotes (page 221)

1. (a) **Operon**: Consists of at least one structural gene coding for the primary enzyme structure, and two regulatory elements: the operator and the promoter.
   (b) **Regulator gene** (repressor gene): Produces a repressor substance that binds to the operator, preventing transcription of the structural genes.
   (c) **Operator**: This is a non-coding sequence of DNA that is the binding site for the repressor molecule.
   (d) **Promoter**: Site of RNA polymerase binding to start the transcription process.
   (e) **Structural genes**: Genes responsible for producing enzymes that control the metabolic pathway.
2. (a) In an inducible enzyme system, the enzymes required for the metabolism of a particular substrate are produced only when the substrate is present. This saves the cell valuable energy in not producing enzymes that have no immediate use.
   (b) Inducible enzyme systems are not adaptive when the substrate is present all (or most) of the time.
   (c) Regulation of a non-inducible system is achieved (in prokaryotes) through **gene repression**: the structural genes are normally transcribed all the time and, when the end product (e.g. tryptophan in *E. coli*) is present in excess to requirements, the genes are switched off.
3. In the inducible system, the genes for metabolizing the substrate are usually switched off, but are switched on when the substrate is present. In the gene repression model, the genes (for metabolizing the substrate) are normally switched on and are only switched off when the substrate is present is excess.

Gene Control in Eukaryotes (page 223)

1. (a) **Promoter**: A DNA sequence where RNA polymerase binds and starts transcription.
   (b) **Transcription factors**: These are proteins that recognize and bind to the promoter sequence and to the enhancer sequence and thereby facilitate initiation of transcription.
   (c) **Enhancer sequence**: The DNA sequence to which the transcription factors called activators bind. This binding is important in bringing the activators in contact with the transcription factors bound to the RNA polymerase at the promoter.
   (d) **RNA polymerase**: The enzyme that, with the initial aid of transcription factors, transcribes the gene.
   (e) **Terminator sequence**: Nucleotide sequences at the end of a gene that function to stop transcription.
2. Difference between gene control in prokaryotes and eukaryotes (any one of):
   - Eukaryotic genes are not found as operons; the control sequences may be some distance from the gene to be transcribed.
   - In eukaryotic gene expression, the transcription factors are important; only when the transcription factors are assembled can the gene be transcribed.
   - Eukaryotic gene expression involves the formation of hairpin loop in the DNA which brings the transcription factors and polymerase into contact.

The Structure of Viruses (page 224)

1. Virion structure: a single type of nucleic acid (DNA or RNA, double or single stranded) and a few enzymes (e.g. reverse transcriptase in HIV), enclosed in a protein coat or capsid. The capsid may be covered by an envelope of lipid, protein, or carbohydrate.
2. Viruses do not conform to any of the criteria by which other life-forms are conventionally classified.
3. They are entirely dependent on using the host's cellular machinery to reproduce.
4. Viruses require living cells in which to replicate and they are very specific to their host. Successful culture requires culture of the right type of cell, and often the cell itself must be in the appropriate physiological state.

Replication in Animal Viruses (page 225)

1. Glycoprotein spikes are important in host recognition and attachment of the virus to the host cell.
2. (a) Endocytosis is the means by which foreign material

is normally engulfed by cells, prior to being destroyed. This response of the cell enables the virus to gain entry into the cell.

(b) Viral DNA replicated in the host cell's nucleus.

(c) Viral proteins synthesised in the host cell's cytoplasm.

3. (a) HIV enters a cells by attaching to the CD4 receptors on a T cell, and fusing with the cell's plasma membrane.
   (b) Reverse transcriptase transcribes the viral RNA into viral DNA. This must occur for the viral genes to be able to integrate into the host's chromosomes where it stays as a provirus.
   (c) The provirus remains integrated with the host chromosome and persists as a latent infection. This means that it can reinfect new host cells whenever the DNA is replicated.

4. (a) Attachment: The virion comes into contact with a cell and adheres to receptor sites on the cell surface. The attachment structures on the viral surface match the receptors on the host cell.
   (b) Penetration: The host cell engulfs the attached viral particle by endocytosis.
   (c) Uncoating: The host's enzymes degrade the protein coat and release the viral nucleic acid into the cell.
   (d) Biosynthesis: The synthesis (assembly) of new infective virions.
   (e) Release: The active virions are budded off from the host cell by exocytosis.

Replication in Bacteriophage (page 226)

1. (a) Lytic cycle: Characterized by multiplication of the virus and lysis of the cell.
   - The virus (or phage) attaches itself to the host cell and inserts its DNA and some enzymes.
   - Virus induces transcription of its own genes and uses cellular machinery to express those genes (produce the viral components).
   - Viral components are assembled and the new (replicated) viruses burst out of the cell (cell lysis).

   (b) Lysogenic cycle (as in λ): Characterized by integration into the host chromosome and cycles of nucleic acid replication (no host cell lysis).
   - The phage attaches to the host cell and inserts its nucleic acid and some enzymes into the host cell.
   - Viral nucleic acid integrates into the host DNA.
   - Viral nucleic acid is replicated (reproduced) along with that of the host cell. This occurs indefinitely until the virus is induced to enter the lytic cycle.

2. Tail region of bacteriophages is important in recognition and attachment to host cell receptor sites.

3. (a) Lysogenic cycle
   (b) Phage DNA has integrated into the bacterial chromosome (and confers the properties encoded by this DNA on the bacterium).
   (c) Properties of virulence or toxin production can be encoded by the viral genes. New virulence may arise in bacteria through the expression of viral genes. Note: There are many examples of this. The pathogenic properties of the bacterium *Corynebacterium diphtheriae*, which causes diphtheria, are related to the production of a toxin encoded by a gene carried by temperate phage (phage in the lysogenic cycle). Similarly, only streptococci carrying a temperate phage can produce the toxin associated with scarlet fever. The botulism toxin produced by *Clostridium botulinum* is encoded by a prophage gene.

KEY TERMS: Mix and Match (page 228)

Anticodon (F), Base pairing rule (BB), Coding strand (CC), Codon (V), DNA Ligase (X), DNA polymerase (O), DNA replication (I), End-product inhibition (U), Exons (Y), Gene expression (C), Gene induction (B), Genetic code (D), Helicase (K), Introns (AA), *Lac* Operon (Q), Lagging strand (G), Leading strand (DD), Lysogenic cycle (L), Lytic cycle (N), Messenger RNA (A), Nucleotides (R), Okazaki fragments (Z), Promoter (P), Replication fork (H), Reverse transcriptase (S), Stop codon (E), Template strand (M), Transcription (J), Translation (W), Transfer RNA (T).

Causes of Mutation (page 230)

1. (a) Radiation: Ultraviolet (UV) rays from the sun, X-rays from medical equipment, gamma-rays from nuclear explosions or nuclear contamination.
   (b) Chemicals: Benzene, formalin (formaldehyde), carbon tetrachloride.

2. A **mutagen** is any physical or chemical agent that increases the frequency of **mutation** (change or disruption in DNA) above the spontaneous (background) rate. Mutagens bring about their effects by disrupting the base sequence of genes, which can affect that gene's product. Mutations to regulatory genes, such as those controlling cell division, are among the most damaging.

3. (a) **Somatic mutations** occur in the body (non-gametic or somatic) cells and are not inherited. They may affect an individual within its lifetime. **Gametic mutations** are mutations to the gametes (in testes or ovaries) and are inherited.
   (b) Gametic mutations are inherited. They are passed on to the next generation and can become part of the genetic variation in the gene pool (upon which natural selection can act).

The Effect of Mutations (page 231)

1. Mutations are the source of new alleles. A neutral mutation has no harmful or beneficial effect under the prevailing conditions, but at some time in the future it may confer some selective advantage (or disadvantage) to the individual that possesses it.

2. Harmful mutations have a detrimental effect on the individual or its offspring, and/or confer a selective disadvantage. For example, the cystic fibrosis mutation and the sickle cell mutation. Beneficial mutations have a beneficial effect on the individual or its offspring, and/ or confer a selective advantage. For example, antibiotic resistance in bacteria.

Gene Mutations (page 232)

1. A **frame shift mutation** occurs when the sequence of bases is offset by one position (by adding or deleting a base). This alters the order in which the bases are

grouped as triplets and can severely alter the amino acid sequence.

2. (a) Reading frame shifts and nonsense substitutions.
   (b) They may cause large scale disruption of the coded instructions for making a protein. Either a completely wrong amino acid sequence for part of the protein or a protein that is partly completed (missing amino acids due to an out-of-place terminator codon).
   (c) A substitution mutation to the third base in a codon. Because of degeneracy in the genetic code, a substitution at the third base position may not change the amino acid that is coded for that position.

3. (a) Mutated DNA: AAA ATA TTT CTC CAA GAT
   mRNA: UUU UAU AAA GAG GUU CUA
   Amino acids: Phe Tyr Lys Glu Val Leu
   (b) ATG → UAC → Tyr = Tyrosine
   (c) No effect because of code degeneracy; both UAC and UAU code for Tyr.

Examples of Gene Mutations (page 233)

1. (b) *Gene name*: HBB *Chromosome*: 11
   Mutation type: autosomal recessive. *May be caused be base deletion, base insertion, or gene deletion (severity depends on the mutation).
   (c) *Gene name*: CFTR *Chromosome*: 7
   Mutation type: autosomal recessive. Great range: deletion, missense, nonsense, misplaced terminator codon. Most common is a deletion of 3 nucleotides.
   (d) *Gene name*: IT15 *Chromosome*: 4
   Mutation type: autosomal dominant. Duplication (CAG repeats of varying length).

2. Few or no β chains causes severe anemia. Frequent blood transfusions are required and this can cause iron accumulation in the tissues and organs.

3. Mutations would have arisen in an individual and spread out gradually from that origin. Thus certain genetic disorders tend to have a higher occurrence in certain regions, especially where the population tends to have stayed relatively isolated geographically.

Cystic Fibrosis Mutation (page 234)

1. (a) mRNA: GGC ACC AUU AAA GAA AAU AUC AUC UUU GGU GGU
   (b) Amino acids: Gly Thr Iso Lys Glu Asn Iso Iso Phe Gly Gly

2. (a) Mutant mRNA: GGC ACC AUU AAA GAA AAU AUC AUC I GGU GGU
   (b) Type of mutation: Triplet deletion.
   (c) Amino acids coded by mutant DNA: Gly Thr Iso Lys Glu Asn Iso Iso I Gly Gly
   (d) Amino acid missing: Phenylalanine (Phe).

3. CF has varying degrees of severity because there are more than 500 mutations of the CF gene. The resulting mutant CFTR protein may not function at all, or it may function only in part, producing a system that is variously effective.

Sickle Cell Mutation (page 235)

1. (a) Bases: 21
   (b) Triplets: 7
   (c) Amino acids coded for: 7

2. **mRNA**: GUG CAC CUG ACU CCU GAG GAG

3. **Amino acids**: Val His Leu Thr Pro Glu Glu

4. **Mutant DNA**: CAC GTG GAC TGA GGA CAC CTC
   Type of mutation: Substitution

5. **Mutant mRNA**: GUG CAC CUG ACU CCU GUG GAG

6. **Amino acids coded by mutant DNA**:
   Val His Leu Thr Pro Val Glu

7. A base substitution causes a change in one amino acid in the hemoglobin molecule. The mutated hemoglobin, being less soluble, causes a distortion of the red blood cells and results in various severe circulatory problems.

Antigenic Variability in Pathogens (page 236)

1. (a) The viral genome is contained on 8 short, loosely connected RNA segments. This enables ready exchange of genes between different viral strains and leads to alteration on the protein composition of the H and N glycoprotein spikes.
   (b) The body's immune system acquires antibodies to the H and N spikes (antigens) on the viral surface, but when different variants arise they are not recognized nor detected by the immune system (there is no immunological memory for the newly appearing antigens).

2. An antigenic shift represents the combination of two or more different viral strains in a new subtype with new properties and no immunological history in the population. Antigenic drifts are much smaller changes that occur continually over time and to which small adjustments are sufficient to provide resistance.

Resistance in Pathogens (page 237)

1. Answer will depend on student choice. All mechanisms are based on mutations that confer greater fitness in the prevailing environment:

- For bacteria: Genes for drug resistance (often carried on plasmid DNA) arise through mutation. In an environment that selects against susceptible strains, resistant bacteria will survive and increase in numbers. Genes for drug resistance are also easily transferred between strains, leading to a spread in resistance while that selection pressure remains.
- For HIV: Resistance may arise though a single mutation or through the accumulation of specific mutations over time. Such mutations may alter the binding capacity of the drug or susceptibility of the virus to the drug. Alternatively, resistance may arise as a result of naturally occurring polymorphisms, which gain favor in the selective environment.
- Chloroquine resistance in *Plasmodium falciparum* is based on the fact that they accumulate significantly less chloroquine than susceptible parasites. The mechanism for this appears to be due to a mutation conferring an enhanced ability to release the chloroquine from the vesicles in which it normally accumulates in the cell.

2. (a-b) any two of the following:
   - Inactivation of the drug.
   - Alteration of the drug's target.
   - Alteration of the permeability of the cell to the drug.

3. Bacteria with several mechanisms of drug resistance can be **difficult to treat** effectively because they have **several ways of withstanding the effects** of a drug on them. The implications of this are vast. Infection and **disease rates increase**, especially within certain populations (e.g. hospitalised patients). The infections can be difficult or impossible to treat with conventional drugs, and if no alternative treatments are available **patients may die.** The potential for drug **resistant bacteria to spread** through the population also increases.

4. Factors contributing to the rapid spread of drug resistance in pathogens include:
   - The typically high mutation rates in viruses, bacteria, and protozoa, combined with short generation times (rapid generational turnover). Note: In bacteria and viruses, high mutation rates are the result of a higher error rate during DNA replication than is typical in most eukaryotic genomes.
   - Strong selection pressures imposed by drug use, coupled with misuse of drugs (e.g. using antibiotics against viral infections), poor patient compliance (patients not taking drugs as prescribed), and the poor quality of available drugs (especially to impoverished populations).

Chromosome Mutations (page 239)

1. (b) **Original**: ABCDEFGHMNOPQRST
   Mutated: AB**FEDC**GHMNOPQRST

   (c) **Original**: 1234567890
   ABCDEFGHMNOPQRST
   Mutated: **ABCDEF**1234567890
   GHMNOPQRST

   (d) **Original**: ABCDEFMNOPQ
   ABCDEFMNOPQ
   Mutated: **ABCDE**ABCDEFMNOPQ
   FMNOPQ

2. Inversion, since there is no immediate potential loss of genes from the chromosome. **Note**: At a later time, inverted genes may be lost from a chromosome during crossing over, due to unequal exchange of segments.

The Fate of Conceptions (page 240)

1. The **maternal age effect** refers to the higher probability of chromosomal disorders in the offspring of older mothers. It is pronounced in the common autosomal trisomies; the chance of producing a child with a trisomic disorder is very low through the early reproductive years (20-35), but increases rapidly after age 40 until the end of a woman's reproductive life (around 50 years).

2. **Amniocentesis** involves extracting a sample of amniotic fluid from the uterus. From a small number of cells floating in the fluid, a karyotype of the baby can be prepared, enabling the detection of many chromosome abnormalities, including common trisomic disorders.

3. Routine tests are now carried out to detect Down syndrome conditions in the fetuses in older mothers. Termination of such pregnancies is increasingly common in some western countries, reducing the incidence in this age group.

Aneuploidy in Humans (page 241)

1. Embryos from left to right: XXY, XO, XXY, XO.

2. (a) Trisomic female (metafemale) (or superfemale)
   (b) Klinefelter syndrome
   (c) Turner syndrome

3. The YO configuration has no X chromosome (the X contains essential genes not found on the Y).

4. (a) For karyotype A: Circle X chromosome
   Chromosome configuration: 45, X (44 + X)
   Sex: female
   Syndrome: Turner
   (b) For karyotype B: Circle XXY chromosomes
   Chromosome configuration: 47, XXY (44 + XXY),
   Sex: male
   Syndrome: Klinefelter

5. Number of Barr bodies:
   (a) Jacob syndrome: 0
   (b) Klinefelter syndrome: 1
   (c) Turner syndrome: 0

6. X chromosome inactivation ensures that the proteins encoded by the genes on the X chromosomes will only be produced by the one active copy.

7. (a) **Nullisomy**: 0, both of a pair of homologous chromosomes are missing.
   (b) **Monosomy**: 1, one chromosome appears instead of the normal two.
   (c) **Trisomy**: 3, three chromosomes appear instead of two, the result of faulty meiosis.
   (d) **Polysomy**: 3+, the condition in which one or more chromosomes are represented more than twice in the cell (includes trisomy).

Trisomy in Human Autosomes (page 243)

1. Autosomal aneuploidies are aneuploidies of the autosomal chromosomes, whereas sex chromosome aneuploidies affect the sex chromosomes (X and Y). (Aneuploidy refers to having a chromosome number that is not an exact multiple of the normal haploid set for the species).

2. (a) Having an extra chromosome may allow for the overproduction of some proteins. Having an extra copy of the gene on the third chromosome may result in more mRNA being produced for that gene.
   (b) **Syndrome**: A suite of symptoms that typically occur together that result from a particular genetic condition. In Down, the syndrome is characterized by a collective suite of abnormalities affecting the face, limbs, internal organs, and musculature.

3. (a) Non-disjunction, which is the failure of chromosome 21 in one of the parents to separate during gamete formation (meiosis). Proportion: 92%.
   (b) Down syndrome phenotype: Mental retardation, retarded growth and short stature, upward slanting eyes, stubby fingers, and folds in the inner corners of the eyes (epicanthic folds). They also tend to

suffer from congenital heart disease.

4. Either one of:
   - Translocation: One parent (a carrier) has chromosome 21 fused to another chromosome (usually number 14). Proportion: less than 5%.
   - Mosaic: Failure of chromosomes 21 to separate in only some cell lines during mitosis (very early in embryonic development). Proportion: less than 3%.

5. 47

KEY TERMS: Word Find (page 244)

```
M I G A M E T I C P T V M N P O L Y S O M Y R A
X T K E V A J X O R G W A P I N E U T R A L N T
F R A M E S H I F T C W S Y N D R O M E H D A J
C X I V D N P R A G Y L O E F J E I E B U E Z E
B Q F A N T I G E N I C D R I F T B A A E H B F
F V C F M T T I Q Y R U R E F G F Q F S A B E A
C S X S Z F J P O L Y P L O I D Y W S V N I J D
D A I R V A N T I G E N I C S H I F T F E U W K
Q O R N N D K Z I N S E R T I O N K F M U G I P
E R G C V Q V K N C B O Y I S D G X W Y P L V X
M Y T M I E N O N D I S J U N C T I O N L U K J
N H B R E N R Y U S Y L P N P W S K K Q O W V P
N C M T I K O S J K P Y O S R S O M A T I C M A
C U F E X S N G I M L H X I B L H P T X D E U R
Z G K V K Z O A E O F Z F X I S F H C V Y L T V
O D Y D R W A M U N N B E B A R R B O D Y B A N
M U T A T I O N Y A I V Y O C C I T F F J V G I
T R A N S L O C A T I O N D U D A C W R Q P E U
Q F J N O C H I M E R A K U G K Y H V P X N N H
K X F G D Q S W C D R C A R R I E R C I R Z Z X
```

Amazing Organisms, Amazing Enzymes (page 246)

1. PCR only works as a viable process when the enzymes being used do not have to be replaced with every cycle. Up until the mid 1980s no enzymes were available that could do this. It was only after *Taq* polymerase was isolated that PCR became a common place technique.

2. *Taq* polymerase is an DNA polymerase enzyme that functions at high temperatures. This means it can remain in the PCR mix during the heating cycles and can be used again during each cycle of DNA replication. Without it, polymerase enzymes must be added after each cycle of heating.

3. Studying the lifestyles of diverse organisms enables us to understand how they cope with environmental challenges. This leads to investigation and isolation of, unique and potentially useful molecules, which can be used in different situations e.g. in industrial processing, in food processing, or medicine.

What is Genetic Modification? (page 247)

1. Organisms may be genetically modified through:
   - The addition of a foreign gene, e.g. human insulin gene inserted into bacteria or yeast for the commercial production of human insulin.
   - Alteration of an existing gene so that a protein is expressed at a higher rate or in a different way. This GM technique is used in gene therapy.
   - Deletion or inactivation of an existing gene, e.g. the Flavr-Savr tomato which has had its ripening gene switched off. Gene inactivation also produces 'knock-out mice' which are used to study the physiological effects of particular genes.

2. (a) **Gene therapy**: A need/desire to find cures/ treatments for genetic diseases (e.g. cystic fibrosis).
   (b) **Transgenic organisms**: A need to accelerate traditional breeding programs by direct manipulation of the genome and a desire to improve the usefulness of livestock and crops by increasing production and reducing susceptibility to diseases and pests. There has also been a desire (need?) to produce new protein products by providing novel genes to make plants and animals into biofactories.
   (c) **Plant micropropagation**: A desire for quick, large scale propagation of plant clones with superior traits, a need for disease free specimens, and a need to overcome seasonal growing restrictions.

Applications of GMOs (page 248)

1. Student's own research activity based on information provided in the student workbook or in other texts.

Restriction Enzymes (page 249)

1. (a) **Restriction enzyme**: Enzymes that cut DNA at very precise base sequences (they are able to create sticky end or blunt end junctions).
   (b) **Recognition site**: The base sequence that a restriction enzyme recognizes and cuts.
   (c) **Sticky end**: The exposed ends of DNA after a restriction enzyme has cut, leaving a partially unmatched base sequence.
   (d) **Blunt end**: The exposed ends of DNA after a restriction enzyme has cut, leaving two blunt ends with no exposed nucleotide bases

2. (a) *Eco*RI
   (b) *Escherichia coli* RY13
   (c) GAATTC

3. (a) GGATCC
   (b) Recognition sites: See below.
   (c) 5 fragments

4. There is a need to have a tool kit of enzymes that allows scientists to cut DNA at any point they wish. The

```
         10         20         30         40         50         60
|AATGGGTACG|CACAGTGGAT|CCACGTAGTA|TGCGATGCGT|AGTGTTTATG|GAGAGAAGAA|
         70         80         90        100        110        120
|AACGCGTCGC|CTTTTATCGA|TGCTGTACGG|ATGCGGAAGT|GGCGATGAGG|ATCCATGCAA|
        130        140        150        160        170        180
|TCGCGGCCGA|TCGXGTAATA|TATCGTGGCT|GCGTTTATTA|TCGTGACTAG|TAGCAGTATG|
        190        200        210        220        230        240
|CGATGTGACT|GATGCTATGC|TGACTATGCT|ATGTTTTTAT|GCTGGATCCA|GCGTAAGCAT|
        250        260        270        280        290        300
|TTCGCTGCGTGGATCCCATA|TCCTTATATG|CATATATTCT|TATACGGATC|GCGCACGTTT|
```


action of such enzymes allows DNA to be manipulated for other recombinant DNA technologies.

Ligation (page 251)

1. (a) **Annealing**: The two single-stranded DNA molecules are recombined into a double-stranded form. Achieved by simple attraction of complementary bases (hydrogen bonds).
   (b) **DNA ligase**: This enzyme joins together the two adjacent pieces of DNA by linking nucleotides in the sticky ends.
2. DNA ligase performs the task of linking together the Okazaki fragments during DNA replication.
3. It joins together DNA molecules, whereas restriction enzymes cut them up.

Gel Electrophoresis (page 252)

1. Purpose of **gel electrophoresis**: To separate mixtures of molecules (proteins, nucleic acids) on the basis of size and other physical properties.
2. (a) The frictional (retarding) force of each fragment's size (larger fragments travel more slowly than smaller ones).
   (b) The strength of the electric field (movement is more rapid in a stronger field).
   Note: The temperature and the ionic strength of the buffer can be varied to optimize separation.
3. The gel is full of pores (holes) through which the fragments must pass. Smaller fragments pass through these pores more easily (with less resistance and therefore faster) than larger ones.

Polymerase Chain Reaction (page 253)

1. PCR produces large quantities of 'cloned' DNA from very small samples. Large quantities are needed for effective analysis. Minute quantities are often unusable.
2. A double stranded DNA is heated to 98°C for 5 min, causing the two strands to separate. Primer starting DNA polymerase are added to the sample. This is then incubated at 60°C for a few minutes during which time, complementary strands are created using each strand of the DNA sample as a template. The process is repeated about 25 times, each time the number of templates doubles over the previous cycle.
3. (a) Forensic samples taken at the scene of a crime (for example, hair, blood, or semen).
   (b) Archeological samples from early human remains.
   (c) Samples taken from the remains of prehistoric organisms preserved in ice, mummified, preserved in amber, tar pits etc.
4. This exercise can be done on a calculator by pressing the 1 button (for the original sample) and then multiplying by 2 repeatedly (to simulate each cycle). Most calculators will not display more than about 8 digits. Alternatively, use a computer spreadsheet.
   (a) 1024 (b) 33 554 432 (33.5 x 106)
5. (a) It would be amplified along with the intended DNA sample, thereby contaminating the sample and rendering it unusable.
   (b) Sources of contamination (any two of):
   Dirty equipment (equipment that has DNA molecules left on it from previous treatments).
   DNA from the technician (dandruff from the technician is a major source of contamination!)
   Spores, viruses and bacteria in the air.
   (c) Precautions to avoid contamination (any two of):
   Using disposable equipment (pipette tips, gloves).
   Wearing a **head cover** (disposable cap).
   Use of **sterile procedures**.
   Use of **plastic disposable tubes with caps** that seal the contents from air contamination.
6. (a) and (b) any of the following procedures require a certain minimum quantity of DNA in order to be useful: DNA sequencing, gene cloning, DNA profiling, transformation, making artificial genes. Descriptions of these procedures are provided in the workbook.

DNA Profiling Using PCR (page 255)

1. **STRs** (microsatellites) are non-coding nucleotide sequences (2-6 base pairs long) that repeat themselves many times over (repeats of up to 100X). The human genome has numerous different STRs; equivalent sequences in different people vary considerably in the numbers of the repeating unit. This property can be used to identify the natural variation found in every person's DNA since every person will have a different combination of STRs of different repeat length, i.e. their own specific genetic profile.
2. (a) **Gel electrophoresis**: Used to separate the DNA fragments (STRs) according to size to create the fingerprint or profile.
   (b) **PCR**: Used to make many copies of the STRs. Only the STR sites are amplified by PCR, because the primers used to initiate the PCR are very specific.
3. (a) Extract the DNA from sample. Treat the tissue with chemicals and enzymes to extract the DNA, which is then separated and purified.
   (b) Amplify the microsatellite using PCR. Primers are used to make large quantities of the STR.
   (c) Run the fragments through a gel to separate them. The resulting pattern represents the STR sizes for that individual (different from that of other people).
4. To ensure that the number of STR sites, when compared, will produce a profile that is effectively unique (different from just about every other individual). It provides a high degree of statistical confidence when a match occurs.

Forensic Applications of DNA Profiling (page 257)

1. Lane A acts a control or calibration lane containing fragments of DNA of known length.
2. Profiles of everyone involved must be completed to compare their DNA to any DNA found at the scene and therefore eliminate (or possibly implicate!) them as suspects.
3. The alleged offender is not guilty. The alleged offender's DNA profile does not appear in the DNA collected at the crime scene nor does it appear in the DNA

database. In fact profile E's DNA is found at the scene.

4. Each individual whale has its own DNA profile. By profiling the whale meat it is possible to identify how many whales were killed by simply counting the number of different profiles. It can then be told if whale meat sold as coming from one whale in fact came from two or more.

Preparing a Gene for Cloning (page 258)

1. **Restriction enzymes** are used to produce the DNA fragment (often a human gene) that is to be cloned, by isolating it from other DNA and providing it with sticky ends. The same enzymes are used to open up the plasmid or viral DNA into which the DNA fragment is to be inserted.
2. (a) Introns are non-protein coding and removing the makes the gene shorter and therefore easier to insert into vectors and easier for bacteria to translate into the protein produce. In the case of PCR, it means large amounts of non-coding DNA are not made.
   (b) Reverse transciptase catalyses the production of a DNA strand from a strand of RNA.
3. Reverse transcriptase is found in retroviruses (such as HIV) and make a copy of DNA from the viral RNA so that the viral genes can be integrated into (and replicated along with) the host's genome.

In vivo Gene Cloning (page 259)

1. *In vivo* methods make it simpler to produce the protein product because the recipient of the gene (the bacterium) will express the gene as its protein product and this product can then be harvested.
2. One would not use bacteria to clone a gene if the purpose of the gene cloning was simply to amplify the DNA, e.g. for forensic or diagnostic purposes, and protein expression not was required. PCR alone is a simple automated process.
3. The gene is prepared by removal of introns. At the same time, an appropriate vector (e.g. plasmid) is isolated. Both the gene and plasmid are treated with the same restriction enzyme to produce identical sticky ends. The DNA fragments are mixed in the presence of DNA ligase and anneal (DNA ligation).
4. 2 x 24 = 48 replications in twenty four hours. Number of bacteria and therefore plasmids = 2^{48} or 2 multiplied by itself 48 times = 281,474,976,710,656 copies.
5. Recombinant colonies can be identified by their ability to grow on agar with ampicillin but not tetracycline.
   - Grow the bacteria on agar containing ampicillin. All resulting colonies must contain the plasmid.
   - Press a sterile filter paper firmly onto the surface of the agar, taking care to mark the paper's position relative to the agar plate. Press the paper onto another agar plate containing tetracycline and mark the position of the paper relative to the plate. Any colonies that do not grow on this new agar must contain the recombinant DNA (as they have been killed by tetracycline).
   - Match the position of colonies between the first and second agar plates. Those on the first plate that are missing from the second are isolated and cultured.
6. A *gfp* marker is preferable over antibiotic resistance genes because antibiotic resistance genes encourage the undesirable spread of antibiotic resistance through bacterial populations. Apart from this, a gfp marker system is much simpler, as the marked colonies can be quickly identified by fluorescence under UV light.
7. Viruses are used in *in vivo* cloning because they are able to insert DNA into a host's genome. Any altered DNA within them will also be inserted and then copied by the cellular machinery of the host.

DNA Chips (page 261)

1. Purpose: To determine the presence or sequence of genes in a sample, and the expression or activity level of those genes.
2. (a) The gene probes making up the microarray fluoresce when cDNA binds to them (called nucleic acid hybridization). This indicates that the RNA product has been expressed in the cell it was taken from. A quantitative amount of gene activity can be computer generated.
   (b) Reverse transcriptase makes a single-stranded copy (cDNA) of the RNA extracted from a cell.
3. (a) Genes that turned red in the microarray (2 and 24) were over-expressed and therefore the most antibiotic resistant genes from the bacteria.
   (b) New antibiotics would be designed to silence these genes that are antibiotic resistant. Alternatively, new antibiotics could also be developed to affect those bacterial genes that are not antibiotic resistant (e.g. genes 4, 17, and 22).
4. They have evolved from Southern Blotting techniques where fragmented DNA is attached to a substrate and then probed with a known gene or fragment. Many different DNA probes are incorporated into a DNA chip. Each spot on the chip has thousands-to-millions of copies of a probe and the color of the spot allows a quantitative read-out of a particular gene's activity in a cell. Advantages include identifying which genes are expressed in a particular tissue and to what extent.
5. The information gained from the microarray could be used to identify which tissue was cancerous and to what degree the cancer was present. This would then enable specialists to advise the best treatment for a particular cancer.

Chymosin Production (page 265)

1. Chymosin (an enzyme) is used to coagulate milk into curds in the production of cheese.
2. Traditional source of chymosin was from the stomachs of young (suckling) calves.
3. (a) Restriction enzymes can be used to cut DNA at specific sites and joined again with DNA ligase. genes of interest can be isolated and inserted into vector DNA e.g. plasmid).
   (b) Bacteria can take up modified plasmid DNA and replicate it in culture.
   (c) The amino acid sequence of a gene protein product

and its mRNA sequence can be determined.
(d) Reverse transciptase can synthesise DNA from mRNA to construct a protein-coding gene.

4. Chymosin gene was isolated by first determining its amino acid sequence and from that, the mRNA coding sequence. Once this is known, a gene probe can be constructed and used to locate the mRNA of interest. Reverse transcriptase is used to produce DNA strand, which can then be amplified using PCR. This technique can be applied to isolating any gene of interest, once its protein product is known.

5. Advantages of using GE chymosin, (a)-(c) any 3 of:
   - The chymosin produced is identical to chymosin from its natural source.
   - The chymosin can be produced without the unnecessary slaughter of calves (a welfare issue for many people) and is suitable for vegetarians.
   - The chymosin is 80-90% active ingredient and is thereby significantly purer as natural rennet, which contains only between 4-8% active chymosin.
   - The chymosin can be produced on demand, in the quantities required. This makes it cost effective.

6. Fungi are eukaryotes. The cells are larger and their secretory pathways are more similar to those of humans than those of *E.coli.*

Golden Rice (page 267)

1. The genes for two different enzymes involved in beta carotene synthesis are taken from two different sources and inserted into the nuclear genome of a rice plant. Expression of the gene under the control of an endosperm specific promoter results in production of beta carotene in the edible portion of the rice plant.

2. The expression of the genes is controlled by a promoter specific to the endosperm, so the genes will only be expressed in that tissue.

3. *Agrobacterium tumefaciens* is a natural plant pathogen and can transfer genes as a consequence of infecting a host plant. The tumour-inducing Ti plasmid can be modified to delete the tumour-forming gene and insert a gene for a desirable trait.

4. (a) Production and consumption of beta-carotene rich rice could alleviate or prevent diseases related to vitamin A deficiency (e.g. night blindness and susceptibility to infection as a result of low immunity). Beta carotene is a precursor to vitamin A.
   (b) Improved nutrition through GM rice will be viable only if the diet in targeted regions is also adequate with respect to fat intake. In some impoverished regions this will not be the case, as diet is inadequate across a wide range of food groups, including fat and protein.

5. (a) More grains per head, larger grains.
   (b) Faster maturation time.
   (c) Improved resistance to pests (e.g. by producing a natural toxin or thicker seed coat etc).

Production of Insulin (page 269)

1. (a) High cost (extraction from tissue is expensive).
   (b) Non-human insulin is different enough from human insulin to cause side effects.
   (c) The extraction methods did not produce pure insulin so the insulin was often contaminated.

2. Mass production of human proteins using *E. coli* facilitates a low cost, reliable supply for consumer use. The insulin protein is free of contaminants and, because it is a human protein, the side effects of its use are minimised.

3. The insulin is synthesised as two (A and B) nucleotide sequences (corresponding to the two polypeptide chains) because a single sequence is too large to be inserted into the bacterial plasmid. Two shorter sequences are small enough to be inserted (separately) into bacterial plasmids.

4. The β-galactosidase gene in *E.coli* controls the transcription of genes, so the synthetic genes must be tied to that gene in order to be transcribed.

5. (a) Insertion of the gene: The yeast plasmid is larger and can accommodate the entire synthetic nucleotide sequence for the A and B chains as one uninterrupted sequence.
   (b) Secretion and purification: Yeast, a eukaryote, has secretory pathways that are more similar to humans than those of a prokaryote and β-galactosidase is not required for gene expression. Secretion of the precursor insulin molecules is therefore less problematic. Purification is simplified because removal of β-galactosidase is not required and the separate protein chains do not need to be joined.

Gene Therapy (page 271)

1. (a) The principle of gene therapy is to correct a genetic disorder of metabolism by correction, replacement, or supplementation of a faulty gene with a corrected version.
   (b) Medical areas where gene therapy might be used: Inherited genetic disorders of metabolism, cancers and other non-infectious acquired diseases, and infectious diseases (e.g. viral infections).

2. Transfection of (and correction of the genes in) **germline cells** allows the genetic changes to be inherited. In this way, a heritable disorder can be corrected so that future generations will not carry the faulty gene(s). Transfection of **somatic cells** only corrects those cells for their lifetime.

3. **Gene amplification** is used to make multiple copies of the normal (corrective) allele.

4. GM stem cells offer longer lasting treatment because the cells will continue proliferating and the therapeutic role will continue as long as the cell line continues. Once a transformed (corrected) somatic cell has reached the end of its life, its therapeutic role ends.

5. (a) Viruses are good vectors because they are adapted to gain entry into a host's cells and integrate their DNA into that of the host.
   (b) Viral vectors can cause problems because (two of):
   - The host can develop a strong immune response to the viral infection. In patients disadvantaged

(immune suppressed) by their disorder, this could severely undermine their health.
- Retroviruses infect only dividing cells.
- Viruses may not survive if attacked by the host's immune system.
- If they do not integrate into the chromosome, the inserted genes may only function sporadically.
- The genes may integrate randomly into chromosomes and disrupt the functioning of normal genes (this occurred recently in retroviral vectors used in SCID gene therapy patients).

6. (a) If a therapeutic gene is integrated into the chromosome, it has a better chance of functioning properly in the long term and being stable (not degraded) in the cell.
   (b) When it integrates into the host's (patient's) chromosome, the gene has the potential to disrupt normally functioning genes (see 1(b) above). **Note**: in recent gene therapy trials in SCID patients, retroviral vectors integrated preferentially into currently active genes.

7. (a) Naked DNA is unstable because it is recognized as foreign and is easily degraded by the normal clean-up mechanisms occurring in the host's tissues (phagocytes etc.). Uptake by cells is inefficient and, once within the cell, the DNA is still at risk of degradation by lysosomes.
   (b) Liposomes offer greater stability because they are formulated so that they are recognized by the host's cell receptors and they target these receptors (so are more directed). They are therefore less likely to be degraded in the tissues.

Gene Delivery Systems (page 273)

1. (a) **CF symptoms**: Disruption of gland function including the pancreas, intestinal glands, biliary tree, sweat glands, and bronchial glands. Infertility occurs in males and females. Disruption of lung function produces the most obvious symptom; the accumulation of thick, sticky mucus in the lungs and associated breathing difficulties.
   (b) CF has been targeted because the majority of cases are the result of a gene defect involving the loss of only one triplet (three nucleotides). In theory, correction of this one gene should not be difficult.
   (c) Correction rate has been low (25%), and the effects of correction have been short lived and the benefits quickly reversed. These problems are related to the poor survival of the viral vector in the body and the sporadic functioning of the gene because it is not integrated into the host's (human) chromosome. Patients suffer problems with immune reaction to the vector. In one patient, treatment was fatal.

2. (a) **Vector**: Adenoviruses.
   Note: potential problems (not required): Gene may function only sporadically when not integrated into the host's chromosome. Adenoviruses have poor survival in the body and are quickly destroyed by the immune system. As a result, corrective rates are low and the effects of the corrective gene are short lived. Patients may show various adverse reactions, as a result of their immune response to the vector.
   (b) **Vector**: Liposomes
   Note: potential problems (not required): Liposomes are less efficient than viruses at transferring genes, so corrective rates are lower than for viral vectors.

3. (a) X-linked SCID is caused by a mutation to a gene on the X chromosome that encodes for the common gamma chain. ADA SCID is caused by a defective gene that codes for the enzyme adenosine deaminase.
   (b) The vector used in the treatment of SCID is a gutted retrovirus (a retrovirus with its natural genetic material removed to make room for the corrective gene being transferred).

4. Gene therapy for cystic fibrosis is targeted at the lungs using either an adenovirus or a liposome delivered via the airways. SCID is treated using a retrovirus introduced to the patient's own bone marrow and the returned to the patient.

5. (a) Chance of interfering with essential gene function: When an essential gene function is affected by gene therapy in somatic cells, the individual will be affected. There may be a chance of corrective therapy in that person's lifetime. When the change affects germline cells, all descendants of the treated individual have a chance to inherit the disrupted gene, so a second heritable defect is created.
   (b) Misuse of the therapy to selectively alter phenotype: Alteration of somatic cells to selectively alter one's phenotype is (presumably) a matter of one's own choice; it may benefit that person in their own lifetime, but will not affect subsequent generations. When these selective changes affect the germline cells, then they are heritable and the alteration is not necessarily limited to one individual. This poses the problem of genetic selection and eugenics, and all their consequent ethical dilemmas.

Investigating Genetic Diversity (page 275)

1. A **proteinase** enzyme dissolves the tissues of the springtails to release the DNA from the cells.

2. (a) *Taq* polymerase binds to the DNA primers and synthesises the complementary strands of DNA.
   (b) Nucleotides are added to the PCR mixture because the polymerase requires free nucleotides to synthesise the complementary DNA strands.
   (c) The high temperature (92°C) separates the strands of DNA. The lower temperature (45°C) causes the primers to anneal to each DNA strand. The sample is then incubated at 72°C because this is the optimum working temperature for the Taq polymerase (a thermophilic bacterial enzyme) which synthesises the complementary DNA strands.

3. (a) The **negative** will be any substance devoid of DNA, e.g. water or some chemical solution blank.
   (b) The **reference** (also called the ladder) provides a reference position for fragments of known length to which the unknown sample fragments can be compared. The fragment length in the sample can then be determined.

4. (a) A point mutation (gene mutation involving a change to one nucleotide).
   (b) The point mutation has occurred to the first base of a codon. Unlike mutations to the third base, which are less likely to change the amino acid coded for, a change to the first base will cause a change in the

amino acid. Another amino acid will form part of the polypeptide chain made from this gene.

The Human Genome Project (page 277)

1. **HGP**: Aims to map the entire base sequence of every chromosome in the human cell (our genome), to identify all genes in the sequence, determine what they express (protein produced), and determine the precise role of every gene on the chromosomes.

2. A HapMap will allow researchers to find genes and genetic variations that affect health and disease. It will also be a powerful tool for studying the genetic factors contributing to variation in response to environmental factors, susceptibility to infection, effectiveness of drugs, and adverse responses to drugs and vaccines.

3. (a) Medical (any of the following):
   - Will identify the location and sequence for up to 4000 known genetic diseases, opening up opportunities for drug therapy.
   - Will provide the information to enable the production of human proteins to correct metabolic deficiencies.
   - Will open up the possibility of gene therapy for many genetic diseases.
   - Will enable the development of new therapeutic drugs to block metabolic pathways.
   - With greater knowledge, emphasis will shift from treatment of disease to better diagnosis and prevention of disease.
   - Screening for genetic predisposition to disease.
   - The ability to sequence quickly and directly will revolutionize mutation research (direct study of the link between mutagens and their effects).

   (b) Non-medical (any of the following):
   - What we learn about human genetics will enable improvement of livestock management.
   - Provides a knowledge base that is a key to understanding the structure, function, and organization of DNA in chromosomes.
   - Provides the basis for comparative studies with other organisms (e.g. for taxonomic purposes).
   - Provides a greater understanding of human evolution and anthropology.
   - Facilitates developments in forensics.

4. **Proteomics** is the study (including identification) of the protein products of identified genes. It relies on the knowledge gained by the HGP, but will ultimately provide the most useful information because it will determine the biological function of the mapped genes.

5. Student's own discussion. Suggestions for each issue listed in the table (pros and cons) are as follows:
   - Rights of third parties:
   (a) They should have the genetic information in order to make an informed decisions (about insurance premiums etc.) to the benefit of those with favorable genetic test results.
   (b) They should be denied the information because they could use it to unfairly discriminate against people with "unfavorable" genetic test results.

   - No treatment, therefore the knowledge is pointless:
   (a) Although there may be no treatment initially, treatment may become available and knowledge of genetic predisposition will allow informed decisions to be made at short notice if necessary.
   (b) Knowing that one has a disease and cannot do anything about it could create emotional problems for many people.

   - High costs of tests:
   (a) Although the costs are high, the knowledge is important to a person's health and to medical research generally and is justifiable.
   (b) If costs are not met by public funds, the high costs will preclude those individuals who cannot personally afford them.

   - Genetic information is hereditary:
   (a) Knowledge of an inherited disease or disorder lets family members assess their risk when planning their own lives (e.g. planning a family).
   (b) Family members may feel forced to not have children if their risk of an inherited disorder is high.

Genome Projects (page 279)

1. (a) Yeast: 461.5 (b) *E. coli*: 957.2 (c) Fruit fly: 93.3 (d) Mouse: 12

2. The amount of (protein)coding DNA ('genes') per Mb of DNA varies tremendously because different species have varying amounts of DNA in non-protein coding regions (regions traditionally not regarded as genes).

3. Sequencing the genomes of major crop plants, such as wheat, rice, and maize, will improve the feasibility of making appropriate, high value, and safe genetic modifications to the plants. **Note**: These modifications may be for characteristics such as improved pest resistance, higher yield, lower water demand, nitrogen fixation etc. Other reasons include a better understanding of crop diseases, growth potential, and genetic resilience in the face of selective (in)breeding.

4. (a) First animal genome sequenced: the nematode worm *Caenorhabditis elegans*
   Date: December 1998 (source Wellcome Trust)
   (b) First plant genome sequenced: Thale cress, *Arabidopsis thaliana*, a weed related to mustard
   Date: December 2000 (source, Nature)

Cloning by Embryo Splitting (page 280)

1. **Embryo splitting** is a much simpler technique than nuclear transfer and involves only the splitting of a normally produced embryo at a very early stage in development. The embryos continue to develop normally (as in the case of natural identical twins). There is no inducement of a somatic cell.

2. (a) Stem cells are undifferentiated; this allows them to be used to make any type of tissue in the recipient (a similar outcome for the production of blood cells is already achieved with bone marrow transplants).
   (b) Cloning high milk yielding cows will enable high yielding herds to be produced quickly (without waiting to see the phenotypic outcome of usual selective breeding processes). Ultimately, this will improve supply at low cost and may also free up land for other uses (since, theoretically, smaller herds would be required).

3. Continued use of embryo splitting will reduce the total pool of genetic diversity from which to select new breeds/strains/varieties.

Cloning by Nuclear Transfer (page 275)

1. **Adult cloning** involves producing genetically identical individuals from the non-embryonic (somatic) tissue of a known phenotype.

2. (a) **Switching off genes in the donor cell**: Induced by low nutrient medium.
   (b) **Fusion of donor and enucleated egg**: Induced by a short electric pulse.
   (c) **Activation of the cloned cell**: Induced by a second gentle electric pulse or by chemical means. A time delay of about 6 hours improves the success of the egg activation process, probably through the prolonged contact of the chromatin with (unknown) cytoplasmic factors.

3. Potential applications (a) and (b) any two of:
   - Production of clones from a proven phenotype that can quickly be disseminated to commercial herds.
   - Rapid production of transgenic animals that produce a particular product (e.g. a pharmaceutical secreted in milk), in order to respond to a market demand.
   - Conservation of rare livestock breeds. It is hoped that cloning will eventually be integrated into zoo management programmes. By retaining the tissues of individuals before they die, some of the genetic diversity of rare species can be retained. It may even be possible to restore species that are on the verge of extinction using cloning technology.

Plant Tissue Culture (page 283)

1. The purpose of tissue culture is to produce large numbers of clones (with identical phenotypic and genetic traits as the parent) in a short space of time.

2. (a) A **callus** is a mass of undifferentiated cells.
   (b) Several plant hormones are added to the culture in sequence. These will stimulate each phase of plant development.

3. Continued culture of a limited number of cloned varieties leads to a change in the genetic composition of the population (the amount of genetic variation decreases). This reduces the amount of variation upon which the gene pool can draw in times of change. This in turn reduces the ability (of the gene pool) to adapt.

4. Compared with traditional propagation methods, tissue culture has a number of advantages and some disadvantages. A discussion could include any of the following points:

   Advantages:
   - Tissue culture enables the production of many clones from a single seed/explant.
   - It allows the selection of desirable traits directly from culture.
   - It facilitates rapid propagation, with no wait for seed production.
   - It is ideal for plants with long generation times, low seed production, or seeds that are difficult to germinate.
   - Tissue culture facilitates the international exchange of plants without quarantine.
   - It allows researchers to eliminate plant diseases from propagation lines.
   - It is also space saving and overcomes seasonal restrictions to propagation.

   Disadvantages
   - Tissue culture is very labor intensive.
   - Trial and error is necessary to determine the ideal culturing conditions.
   - Cultured plants may be genetically unstable/infertile and, over time, there may be a loss of genetic diversity (see Q3.).

The Ethics of GMO Technology (page 279)

1. Plants produce pollen which has the potential to be spread in a broadcast fashion (broadcast pollination). This increases the risk that genes (e.g. for herbicide resistance) will be transferred from a GM plant to a weed or other plant. **Note**: Such gene transference has already been demonstrated between plant species. Transfer of sex cells (and therefore genes) between animals in this way does not occur; breeding in animals is generally a more precise and difficult process.

2. (a) **Advantage**: Crop growers could spray a field with herbicide to kill weeds without harming the crop.
   (b) **Problem**: Herbicide resistance may be spread to weed plants by viral or bacterial vectors that infect plants. Encourages overuse of herbicide chemicals.

3. (a) **Advantage**: Ability to grow tropical food crops in regions that could not previously do so.
   (b) **Problem**: Such plants in a new environment may become pest species. Undeveloped economies that rely on tropical cash crops may suffer as a result of competition from economically strong countries.

4. (a) **Advantage**: Will allow regions that are poor in agricultural production to produce crops.
   (b) **Problem**: Such plants in a new environment may become pest species. Disturbs natural wetland habitats, probably resulting in the loss of wetland and marsh native species.

5. (a) Enhancing wool production in sheep (yield and/or wool quality).
   (b) Use of livestock animals as biofactories by producing useful proteins in their milk (especially cattle, but also sheep and goats).

6. The widespread use of antibiotic markers in food crops for human consumption or stock food may give rise to antibiotic resistant strains of pathogenic bacteria which affect humans and stock animals. Restrained use of antibiotics is now considered essential in preventing large scale development of antibiotic resistance.

7. (a) Introduces nitrogen fixing ability in non-legumes thereby reducing the need for nitrogen fertilizers.
   (b) The bacterium would prevent attack on the seeds by pathogenic bacteria and fungi.

8. (a) Some points for discussion are:
   - That the GM product and/or the GMO could have some unwanted harmful effect on humans or other organisms.
   - That the genetic modification would spread uncontrollably into other organisms (breeding populations of the same or different species).
   - Consumer choice is denied unless adequate labeling protocols are in place. If everything contains GM products then there is no choice.
   - General fear of what is not understood (fear of

real or imagined consequences).
- Objections on the grounds that it is ethically and morally wrong to tamper with the genetic make-up of an organism.
- Generation of monopolies where large companies control the rights to seed supplies and breeding stock.

(b) Those that pose a real biological threat are:
- Amongst plant GMOs, the indiscriminate spread of foreign genes.
- Unusual physiological reactions e.g. allergies, to novel proteins.
- Some animal rights issues may be justified if genetic modification causes impaired health. **Note**: This question is not intended to imply that ethical or moral concerns are less valid than biological ones. It is merely an exercise in identifying the nature of the biological concerns.

KEY TERMS: Mix and Match (page 287)

Amplification (J), Annealing (B), Blunt end (U), DNA chip (N), DNA ligation (V), DNA marker (F), DNA polymerase (O), DNA profiling (C), Gel electrophoresis (Q), Gene technology (R), GMO (P), Marker gene (X), mtDNA (W), Microsatellite (M), Model organism (G), Molecular clone (Y), PCR (H), Plasmid (I), Primer (D), Recognition site (S), Recombinant DNA (K), Restriction enzyme (E), Sticky end (T), Transformation (L), Vector (A).

Life in the Universe (page 290)

1. The study of comets may provide us with an understanding of how the first organic molecules were formed and whether or not comets could have carried them to Earth.

2. Capsule-like droplets could have provided an appropriate environment for the enclosure of prebiotic molecules (just like plasma membranes).

3. Evidence from spectral analysis of the light received from stars and from examination of meteorites.

The Origin of Life on Earth (page 291)

1. (a) **Ocean surface**: Suggests that life arose in a tide pool, pond, or moist clay. UV/lightning would have energized volcanic gases to form the prebiotic molecules in froth.
   (b) **Panspermia**: Proposes that living organisms were seeded on Earth from comets and meteors.
   (c) **Undersea thermal vents**: Proposes that life arose at volcanic vents. The vents provided the appropriate environment (anoxic, with the necessary gases, energy, and catalysts), for the formation of prebiotic compounds.

2. Ribozymes are able to act both as genes **and** enzymes. This enabled a plausible model for the origin of life to be constructed because the ribozyme (RNA) molecules could perform the catalytic activity necessary to assemble themselves. Once formed, they could synthesize further proteins. Without a molecule (ribozyme) with this dual ability, genes could not be formed without enzymes and enzymes (proteins) could not be formed without genes.

3. 3.5 billion years old (cyanobacteria)

4. (a) Any of: Mars, Europa (moon of Jupiter), Titan (moon of Saturn).
   (b) Finding life elsewhere in our solar system could suggest that life on Earth was seeded from other regions in space (it did not have to arise on Earth itself). At the very least, it would show that the origin of life on Earth was not unique.

Prebiotic Experiments (page 293)

1. (a) Primeval atmosphere: Reaction chamber.
   (b) Primeval ocean: Flask containing boiling water.
   (c) Lightning: Electric discharge from power supply (7500 V, 30 A) to two tungsten electrodes.
   (d) Volcanic heat: Heating element below the flask containing the water.

2. Amino acids (in later experiments, also nucleic acids, sugars, lipids, adenine, ATP).

3. (a) Possible reason: The early atmosphere may have had a different composition to that produced in the experiment, i.e. CO, CO_2, N_2 (rather than methane). **Note**: In this case, it has been suggested that carbon atoms will not break out to form larger, organic molecules as Miller and Urey suggested.
   (b) The conditions of the experiment could be adjusted with the addition of CO_2, CO, and N_2. Note: In recent experiments with CO_2 present, using similar apparatus, the experimental outcome was virtually unchanged in terms of the compounds produced. However, yields were substantially lower.

The Origin of Eukaryotes (page 294)

1. In sequence 1, a cell with a symbiotic "pre-mitochondrion" led to present-day animal cells. Such a cell then also engulfed a photosynthetic prokaryote (pre-chloroplast), leading to present-day plant cells. In sequence 2, a plant-like cell containing both aerobic and photosynthetic symbionts could have later lost its chloroplast symbiont, giving rise to animal-like cells.

2. (a) Mitochondria: Originally an endosymbiosis between an aerobic (purple) bacterium and an early cell. The bacterium was engulfed by a nucleated pre-prokaryote and became an "pre-mitochondrion". **Note**: The symbiosis would be progressive, with the aerobic bacterium relying more and more on the host cell for materials other than energy. The host cell would rely increasingly on the pre-mitochondrion for energy.
   (b) Chloroplasts: Originally an endosymbiosis between a photosynthetic bacterium and an early cell. The bacterium (similar to cyanobacteria) was engulfed by a nucleated pre-prokaryote and became an "pre-chloroplast". **Note**: As above, the symbiosis would become progressive. The host cell may or may not have already had a pre-mitochondrial structure.

3. Both mitochondria and chloroplasts retain their own self-replicating circular chromosome (DNA evidence of a prokaryote origin). The DNA has a code that is identical to that used by prokaryotes. Chloroplasts have an internal membrane structure that is nearly identical

to that of modern cyanobacteria.

4. Fossil evidence supports the endosymbiont theory as the first firm evidence of eukaryote cells appears in the fossil record at 540-600 million years ago. Cyanobacterial fossils (photosynthetic bacteria) appear in the fossil record much earlier than this.

The History of Life on Earth (page 295)

1. Animal cells rely on aerobic respiration and so require free oxygen in order to carry out their metabolic processes. **Note**: Complex animals have a greater oxygen requirement (demand) than simple animals, therefore only once oxygen was readily (freely) available in the atmosphere, could complex animal life forms evolve.

2. (a) Invertebrates: 550 mya (d) Reptiles: 265 mya
(b) Ray-finned fish: 400 mya (e) Birds: 185 mya
(c) Land plants: 375 mya (f) Mammals: 205 mya

3. (a) Mass extinctions result in the vacation of many niches (as their occupants perish). This provides new opportunities for survivors to undergo adaptive radiation and diversify to occupy the vacant niches.
(b) Initially, any diversity present on Earth was simply chemical diversity. With the origins of life, but before the development of eukaryote cells, life forms were all rather similar (simple cells). With the development of eukaryotes, and then multicellularity, there existed the potential for the development of complex organisms and an explosion in the diversity of life forms. Many of the organisms on Earth today have their origins in the Precambrian and early Paleozoic explosion of life.

Fossil Formation (page 297)

1. (a) **Pyritization**: Iron pyrite replaces hard remains of the dead organisms.
(b) **Amber**: Conifer resin or gum traps insects or other small invertebrates and then hardens.
(c) **Petrification**: Wood is silicified: silica from weathered volcanic ash is incorporated into the decayed wood.
(d) **Phosphatization**: Bones and teeth are preserved in phosphate deposits.
(e) **Tar pit**: Organisms fall into a tar pit (mix of sand and tar) and are trapped there. Their remains become embedded in the matrix of sand and tar.

2. Decay.

3. **Transitional fossils** are those possessing a mixture of the characteristics of two different, but related, taxonomic groups. They are important because they indicate that one group may have given rise to the other by evolutionary processes.

The Fossil Record (page 299)

1. (a) Any one of: Horse, elephant, pig, numerous dinosaur groups, trilobites.
(b) Any one of: Tuatara, coelacanth, ginkgoes (ancient conifers).

2. Fossilized plants and animals from the fossil record can be compared to living species today thus giving an insight into how they have evolved over time.

3. (a) Layer A (b) Layer H (c) Layer I (d) Layer O

4. (a) Layer E
(b) It has the same relative position in the sequence of layers and it contains fossils typical of the layer.

5. These rocks may be so old that large organisms with hard body parts were not present to be fossilized (i.e. they may have had soft body parts that decomposed).

6. (a) Layers C and F (b) Layer J

7. Any three of:
Radiometric dating: Measuring radio-isotope ratios.
Fossil correlation: Matching up fossil community types with those at another location with a known date.
Paleomagnetism: Measuring the magnetic alignment of the rock and correlating this with known magnetic pole directs in the Earth's past.
Fission track analysis: Measuring the number of tracks caused by particles in rock crystals.

8. (a) Layer A: 0 – 80 million years old (< 80 mya)
(b) Layer C: 80 – 270 million years old
(c) Layer E: 270 – 375 million years old
(d) Layer G: Older than 375 million years old
(e) Layer L: 270 million years old
(f) Layer O: Older than 375 million years old

Dating a Fossil Site (page 301)

1. **Occupation horizons** (layers in the soil profile with evidence of human occupation) are indicators of human activity some time in the past. Each one is the ancient living floor of the site and can provide information about the lifestyle and culture of the human occupants.

2. (a) Older than 18 500 but younger than 45 000 years.
(b) The upper surface is about 18 500 years since it has the hearth (which has been dated) near the top.

3. (a) Pottery bowl: Thermoluminescence.
(b) Skull: Radiocarbon-14, Uranium-thorium, Electron spin resonance.
(c) Hearth: Radiocarbon-14, Thermoluminescence.
(d) Tooth: Electron spin resonance.

4. Paleoanthropologists document **all** of the remains at a fossil site to maximize the recovery of information. This allows for cross referencing between all finds, i.e. of hominin and non-hominin species, and gives a more accurate picture of the past environment and ecology.

5. Paleoanthropology is a multidisciplinary science encompassing paleontology, geology, prehistoric archeology, molecular biology, and behavioral science amongst others. Different scientific disciplines provide information on different aspects of a site's physical and biological environment and allow any site of hominin activity to be examined in a holistic way. Working in this way, scientists can piece together a more accurate interpretation of past ecology and behavior.

Protein Homologies (page 303)

1. Chimpanzees and gorillas have virtually identical amino acid sequences to humans for some proteins.

2. (a) They play a crucial role in the respiratory pathway.

Most changes are likely to be deleterious so they change very little over time.

(b) Such proteins are good candidates for use in establishing homologies because the few changes that are retained through time are likely to be meaningful, i.e. represent major divergences in evolutionary lines.

3. Any of: The rate of change must be calibrated against material evidence (e.g. fossils) for firm conclusions to be made. The functions of the protein may change over time. Clock may run at a different rate in different species.

DNA Homologies (page 305)

1. The similarity of DNA from different species can be established in a rudimentary way by measuring how closely single strands from each species mesh together. The more similar the DNA, the harder it is to separate them. These studies have confirmed most evolutionary relationships guessed at from anatomical comparisons.
2. (a) Chimpanzee (b) Galago
3. (a) 7 – 8 (b) Approx. 12
4. 45 million years ago.

Comparative Anatomy (page 306)

1. (b) Human arm: Modified for tree climbing (flexible joints) and improved dexterity of fingers.
   (c) Seal's flipper: Modified to increase surface area and streamlined to function as a paddle.
   (d) Dog's foot: Modified for swift running in pursuit of prey. Walks on toes, long limbs to provide lengthened, running stride.
   (e) Mole's forelimb: Short and strong limb. Shovel-like paw with sharp claws for digging and propelling itself underground.
   (f) Bat's wing: Modified into a wing for flying. Very long metacarpals and fingers stretch the skin into a wing.
2. The limbs all share the same basic bone anatomy, although highly modified in some cases. It is possible to match bone for bone but, at the same time, recognize how individual bones or bone groups have changed to better perform a new function for the animal.
3. Innate or genetically determined behavior is inherited in the same way as structural features. While some behavior is learned, this tends to occur within, rather than between species.

Vestigial Organs (page 307)

1. When an organism adopts a new niche, exploits a new habitat, or takes on 'new' way of doing something, some existing structure may become redundant. Rather than helping the organism to exploit its new way, they may hinder it. Selection pressures may then act against the organs being large. Even if there is no direct selection pressure, the organ may still regress with time as less energy is invested in a little used body part.
2. The genes that code for it are still present and continue to express themselves to produce the structure. What is required (for its loss) is an accumulation of mutations that will cause the relevant genes to be switched off.
3. It is possible to see a gradual reduction in the size of the vestigial organ from the early ancestor, through transitional forms, to modern forms.

Evolution of Horses (page 308)

1. (a) **Teeth**: The equid tooth structure changed from a generalised omnivorous tooth to a more specialised herbivorous tooth designed to cope with the tough grasses which became their primary diet. Changes included elongation of the teeth, development of a flat grinding surface, and a hard cement like covering to protect the tooth and cope with the harsh diet.
   (b) **Limb length**: The grasslands were a more open environment which made the equids more susceptible to predation. The increasing limb length enabled the equids to develop a better view of the land around them and thus spot predators, and also to run more quickly.
   (c) **Toe number**: The number toes reduced from four to one and the equids began to stand on tiptoe. These developments promoted and efficient forward and back stride which enabled the equids to become specialised runners (cursors), giving them an advantage in the wide open plains which were developing at the time.
2. The evolution of the modern horse is well documented in the fossil record with numerous fossils, including transitional fossils which show the major features of equid evolution over time (change in body size, limb length, tooth structure, toe reduction). The extensive fossil record has enabled scientists to put together a comprehensive and robust model of horse evolution (including speciation and extinctions).

The Evolution of Novel Forms (page 309)

1. **Evo-devo**, or evolutionary developmental biology, is the study of the origin and evolution of embryonic development in animals. It traces the role of genes in developmental processes and looks at how modifications of these processes may lead to the evolution of novel features.
2. Homeotic gene sequences, such as the *Hox* genes, make up a genetic tool kit for animal development. These genes act as genetic switches (transcription factors), turning other genes on or off in the course of development. For example, the *Hox* genes determine the positioning of body parts in organisms as diverse as *Drosophila* and mice. The same genes are present in essentially all animals, including humans, and mutations in these genes can have a profound effect on morphology.
3. Evo-devo is important as evidence for evolution because it explains how novel forms may arise without attributing these changes to the evolution of new genes. Instead natural selection associated with gene regulation can be explained as a mechanism for the evolution of novel forms.
4. Changes in gene expression can bring about changes

in morphology, e.g. in differences in neck length in vertebrates and eyespot development on butterfly wings. Neck length is controlled by the expression of the *Hox c6* gene and where it is expressed marks the boundary between neck and trunk vertebrae. For example, in snakes the boundary is shifted forward to the base of the skull with no neck results, while in geese the boundary is shifted backwards and a long neck results. Similarly, the development of eyespots on the wings of butterflies is controlled by switches in the *Distal-less* gene, whose expression results in a range of spots from virtually all eyespot elements expressed to very few.

Oceanic Island Colonizers (page 311)

1. The plants and animals have arrived on these islands from the nearby mainland communities (South America and Africa). There is no effective way plants and animals could move between these two island groups.

2. Trade winds and ocean currents predominantly move from west to east. This makes traveling from South America to Tristan da Cunha easier than travelling from Africa to Tristan da Cunha.

3. (a) A large island area provides (normally) a greater variety of habitats (and potential niches) for colonizers to exploit, increasing the chances of adaptive radiation and a greater biodiversity.
   (b) The longer an island is isolated from the mainland the greater chance the species there will become unique (either by their own genes changing or the genes of their ancestors on the mainland changing) as there is no exchange of genes between isolated species and their original gene pool.
   (c) Islands that are relatively close to a continental land mass are likely to have a biota that is similar to that of the mainland (because of regular exchanges of genes). The island biota will be unlikely to have the same degree of endemism as an island further from a continental land mass.

4. Features might include:
   - A marine habit: Colonizers must be able to tolerate salt water in order to reach isolated islands.
   - Able to fly but not necessarily strongly: Small birds and insects may be carried on wind currents/blown to an island and then remain there (not strong enough to make a return flight). Stronger fliers, e.g. marine birds, can easily reach offshore islands but are strong fliers and don't remain there.
   - Able to survive long periods without food or water: Isolated island (not surprisingly) are often hundreds to kilometers from land. Travelling to them on rafts of drift wood can take many weeks. Animals with a metabolism that can slow significantly (e.g. reptiles) are more likely to survive the journey.

Continental Drift and Evolution (page 313)

1. South America, Antarctica, Africa, Madagascar, India, Australia, New Guinea, New Caledonia, New Zealand.

2. & 3. See diagram below.

4. Once the continents have been fitted together correctly, the direction of the polar regions for each continent match in a way that shows that the continents were grouped near the South Pole in the past.

5. *Lystrosaurus* was distributed throughout much of Gondwana **before** the supercontinent broke up.

6. (a) Africa and India broke away early from Gondwana, before the first appearance of *Nothofagus*.

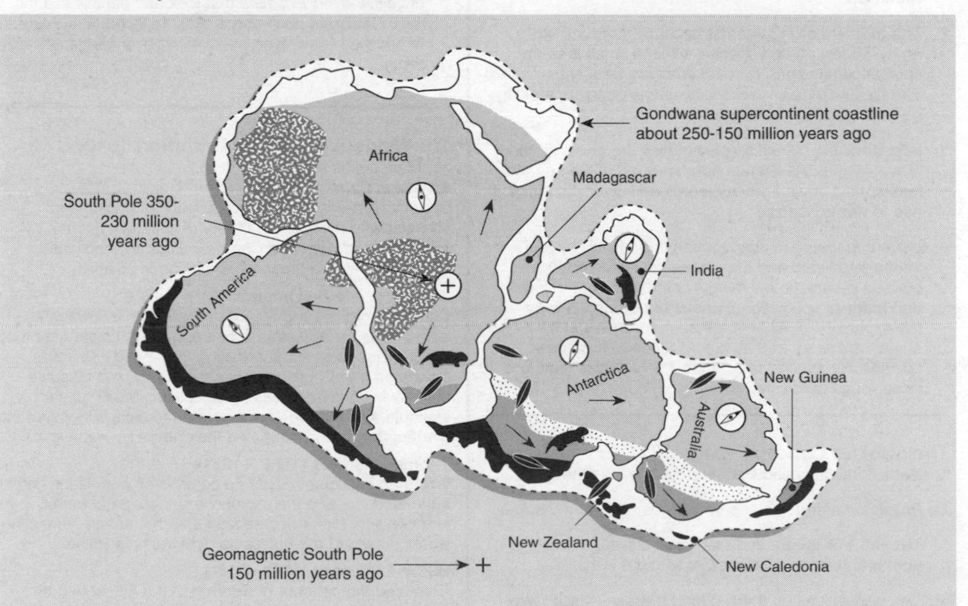

(b) Southern beech distribution:

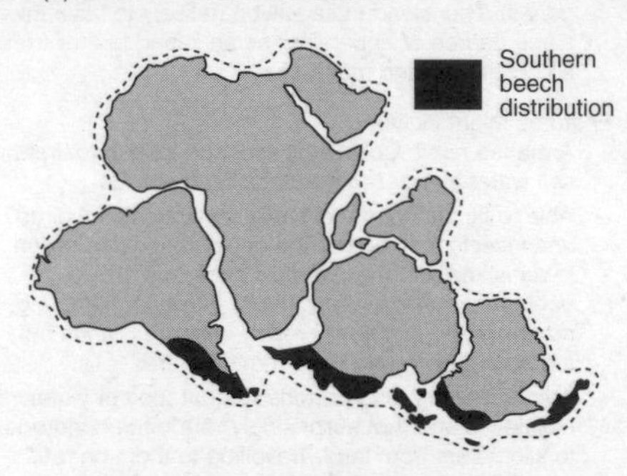

(c) They match up in a distribution pattern that is consistent with their being widespread in Gondwana after the separation of India and Africa.

7. Rate of 2 cm per year = 1 m per 50 years
= 1 km per 50 000 years
= 2300 km per 115 million years

8. Continental drift provides a good explanation for the current distribution patterns of existing and fossil species. In fact, the distribution patterns only make sense if continental drift is accepted. It also provides an indication of the long periods of time required for the patterns to be achieved.

Small Flies and Giant Buttercups (page 320)

1. When the original species of drosophilidae arrived on the Hawaiian islands it found many new unoccupied niches into which it expanded, leading to an **adaptive radiation**.

2. The fruit flies are of interest because there are so many closely related species within a small area an speciation has been (and is) relatively frequent. The flies also have a relatively simple genome, making genetic studies relatively easy.

3. In general the oldest species of flies are found on the oldest islands. As islands appeared out of the sea the flies spread to new environments and diversified, giving rise to newer species.

4. Buttercups living in alpine areas periodically have their habitats reduced and their range restricted during periods of climatic warming. This restricts gene flow and leads to speciation. Periods of cooling allow for the expansion of their range and movement to new environments as well as hybridization to form new species. Repeated many times, these cycles lead to a large range of species.

The Species Concept (page 321)

1. Behavioral (they show no interest in each other).

2. Physical barrier; sea separating Australia from SE Asia.

3. The red wolf is rare and may have difficulty finding another member of its species to mate with.

4. The populations on the two land masses, which have identical appearance and habitat requirements, were connected relatively recently by a land bridge during the last ice age (about 18 000 years ago). This would have permitted breeding between the populations. Individuals from the current populations have been brought together and are able to interbreed and produce fertile offspring.

5. Several definitions of a biological species are possible. Most simply, a species is the lowest taxonomic grouping of organisms. From a functional point of view, a species is a group of organisms that are freely interbreeding (or potentially so) but reproductively isolated from other such groups. Species are usually (but now not exclusively) recognized by their morphological characters. Cryptic species are morphologically indistinguishable, but reproductively isolated as a result of behavioral or other differences. "Species" that do not reproduce sexually (with the combination of gametes) at any stage provide problems for a standard species definition. Such organisms include bacteria, where genera and 'species' are distinguished on the basis of structure and metabolism. In many cases, species are really types or strains.

6. (a) There is some gene flow between neighboring subspecies indicating they are not genetically isolated.
(b) Where the two ends of a species population overlap any interbreeding does not result in fertile hybrids; the gene pools are isolated.

7. The selection pressures on yellow-eyed *Ensatina* (presumably the benefits accrued by avoiding predation) must be similar to those the toxic California newt, and stronger than selection pressures favoring continued breeding with the Sierra Nevada *Ensatina* populations. Directional selection is favoring individuals of yellow-eyed *Ensatina* that look and behave like the toxic California newt and is likely to increase the rate at which the yellow-eyed population becomes genetically distinct.

The Modern Theory of Evolution (page 323)

1. Main points are:

Erasmus Darwin (1731 – 1802):
Charles Darwin's grandfather. English physician and scientist who proposed a theory of evolution by the *Inheritance of Acquired Characteristics*. Probably had a great influence on the thinking of Charles Darwin.

John Baptiste de Lamarck (1744 – 1829):
Contributed greatly to the classification of invertebrates. Best known for a mistranslation of his 1809 statement that species had evolved by adapting to a "need". This was translated into English as "desire" and he was ridiculed unjustly by English speaking scientists. (Much quoted example was that he was supposed to have suggested that giraffes could have stretched their necks by *wanting* to).

Thomas Malthus (1766 – 1834):
Wrote an important essay on controls on population growth that helped inspire the evolutionary theories of both Darwin and Wallace. Malthus proposed that the human population would be wiped out unless its birthrate was limited.

Herbert Spencer (1820 – 1903):
Proposed the concept of Survival of the Fittest that he

first used in his book *Principles of Biology* (1864). This idea was adopted by Darwin and Wallace in their theory of *Evolution by Natural Selection*.

Charles Lyell (1797 – 1875):
British geologist famous for promoting the work of James Hutton of *Geological Uniformitarianism* (that the same agencies are at work in nature today, operating at the same intensities as they have always done throughout geological time, e.g. erosion and sedimentation rates).

Alfred Russel Wallace (1823 – 1913):
Wallace jointly proposed the theory of evolution by natural selection with Darwin. One of the first people to map the distribution of living things leading him to propose the world is divided up into biogeographical zones. He wrote to Darwin of his ideas on evolution by natural selection, spurring Darwin on to publish *The Origin of Species*.

Gregor Mendel (1822 – 1884):
Explained the process of inheritance involving "particles" that are passed on undamaged through the sex cells to the next generation. This was a radical departure from the view of the day that inheritance involved a mere *blending* of parental characteristics.

August Weismann (1834 – 1914):
German biologist who is regarded as the father of modern genetics. He discredited the theory that acquired characteristics could be inherited. He was the first to propose that chromosomes are the basis of heredity.

Theodosius Dobzhansky (1900 – 1975):
A Russian biologist working in the area of population genetics. One of the architects of the Modern Synthetic Theory of evolution. Published *Genetics and the Origins of Species* in 1937 - a publication that marked the beginning of a new understanding of evolutionary biology. Dobzhansky studied isolating mechanisms and showed how speciation could occur.

Ernst Mayr (born 1904 – 2005):
A German evolutionary biologist who collaborated with Dobzhansky, Julian Huxley, and George Gaylord Simpson to formulate the modern evolutionary synthesis. Worked on speciation in animals and defined different types of speciation mechanisms. Also proposed in the 1950s that rapid speciation events could occur. This became important for later ideas on punctuated equilibrium.

Julian Huxley (1887 – 1975):
English evolutionary biologist who worked on ritualization behavior, neoteny, and allometric growth before collaborating with Dobzhansky, Mayr, and George Gaylord Simpson to formulate the modern evolutionary synthesis. Wrote: Evolution: *The Modern Synthesis* in 1942 after which the new theory was named.

J.B.S. Haldane (1892 – 1964):
English geneticist who contributed to the Modern Synthetic Theory of evolution. Remembered most as an innovative pioneer in population genetics, a field which reshaped modern evolutionary biology.

Sewall Wright (1889 – 1988):
American population theorist and one of the architects of the new evolutionary understanding. Together with Haldane and Fisher, Wright gave evolutionary biology a mathematical basis by working out the mathematical principles of population genetics. This transformed Darwinism into a 20th century science. Best known for his contributions to knowledge of evolution in small populations and the Founder effect (called the Sewall-Wright Effect).

R. A. Fisher (Sir) (1890 – 1962):
English statistician and geneticist who contributed to the Modern Synthetic Theory of evolution. He showed that Mendel's work and Darwin's ideas on natural selection were in agreement, not conflict as some had believed. Made major contributions also to the development of statistical ideas and to knowledge of human inheritance. Note that although Haldane, Fisher, and Wright made related contributions to the knowledge of population genetics, they were not collaborators and did not view themselves as such.

Darwin's Theory (page 324)

1. **Natural selection** can provide the means for species change over time because natural selection will always favor the most adaptive phenotypes (therefore genotypes) at the time. More favorable phenotypes will have greater reproductive success and will become proportionally more abundant in the population. Over time, favorable phenotypes will predominate and the unfavorable phenotypes will become very rare.

Adaptations and Fitness (page 325)

1. **Adaptive features** are genetically determined traits that have a function to the organism in its environment. Physiological 'adaptation' (**acclimatization**) refers to the changes made by an organism during its lifetime to environmental conditions (note that some adaptive features do involve changes in physiology).

2. Shorter extremities are associated with colder climates, whereas elongated extremities are associated with warmer climates. The differences are associated with heat conservation (shorter limbs/ears lose less heat to the environment).

3. Large body sizes conserve more heat and have more heat producing mass relative to the surface area over which heat is lost.

Natural Selection (page 326)

1. Environmental instability provides a greater chance that extreme phenotypes might be advantageous, for example in being able to exploit newly available food sources or habitats. In contrast, the predictability of a stable environment encourages a reduction in phenotypic variability (narrow phenotype suited to the prevailing, stable environment).

2. (a) Drought favored birds with beak sizes at two extremes of the range, since these birds could exploit the small and the large seed sizes.
   (b) Continuation of the drought could lead to further divergence in beak size.
   (c) The selection pressures favoring a bimodal distribution in beak size would be reduced and medium sized beaks would become relatively more common (i.e. there would be a shift from disruptive to stabilizing selection).

Selection for Skin Color in Humans (page 327)

1. (a) Folate is essential for healthy neural development. Note: A deficiency causes (usually fatal) neural tube defects (e.g. spina bifida).
   (b) Vitamin D is required for the absorption of dietary calcium and normal skeletal development. Note: A deficiency causes rickets in children or osteomalacia in adults. Osteomalacia in pregnancy can lead to pelvic fractures and inability to carry a pregnancy to term.

2. (a) Skin cancer normally develops after reproductive age and therefore protection against it provides no reproductive advantage and so no mechanism for selection.
   (b) The new hypothesis for the evolution of skin color links the skin color-UV correlation to evolutionary fitness (reproductive success). Skin needs to be dark enough to protect folate stores from destruction by UV and so guard against fatal neural defects in the offspring. However it also needs to be light enough to allow enough UV to pentrate the skin on order to manufacture vitamin D for calcium absorption. Without this, the female skeleton cannot successfully support a pregnancy. Because these pressures act on individuals both before and during reproductive age they provide a mechanism for selection. The balance of opposing selective pressures determines eventual skin coloration.

3. Women have a higher requirement for calcium during pregnancy and lactation. Calcium absorption is dependent on vitamin D, making selection pressure on females for lighter skins greater than for males.

4. The Inuit people have such abundant vitamin D in their diet that the selection pressure for lighter skin (for UV absorption and vitamin D synthesis) is reduced and their skin can be darker.

5. (a) Higher chances of getting rickets or (the adult equivalent) osteomalacia due to low UV absorption.
   (b) The simplest option to avoid these problems is for these people to take dietary supplements to increase the amount of vitamin D they obtain.

Industrial Melanism (page 329)

1. The appearance of the wings and body (how speckled and how dark the pigmentation).

2. (a) Selective agent is selective predation by birds.
   (b) The selection pressure (the differential effect of selective predation on survival) changed from favoring the survival of light colored morphs in the unpolluted environments prior to the Industrial Revolution, to favoring the dark morph (over the light morph) during the Industrial Revolution (when there was a lot of soot pollution). In more recent times, with air quality improving, the survival of the light colored morphs has once again improved.

3. There was a high frequency of the melanic form in and around industrial regions; the grey mottled form was more prevalent in non-industrial regions.

4. (a) Summer smoke from 75 to 15 μgm^{-3} (80% drop). Winter SO_2 from 140 to 15 μgm^{-3} (90% drop).
   (b) Melanic form frequency dropped from 95% to 50%

5. In specific regions, the selection was directional, although over all of England, it could be considered disruptive (more correctly balancing), because different phenotypic extremes are favored in different regions.

6. **Natural** selection occurs without human intervention, in response to normal environmental changes. In contrast, **artificial** selection is directed specifically by humans, usually towards a specific phenotypic result.

7. The environment (including the biotic environment) determines the selection pressures to which an organism will be exposed. It is these selection pressures that determine the fitness (differential survival and reproduction) or otherwise of an organism (with its particular set of allele combinations).

Heterozygous Advantage (page 331)

1. People who are heterozygous for the sickle cell gene are somewhat affected by sickle cell anemia but have considerable resistance to malaria which is widespread in the region. This **heterozygous advantage** maintains the mutant allele at a relatively stable frequency in the population despite its deleterious effects. The stable coexistence of both the sickle cell allele and the normal allele exemplifies a **balanced polymorphism**.

Selection For Human Birth Weight (page 332)

Sample data and graph below. Note: For the construction of weight classes, it is necessary to have a range of weight categories that do not overlap. The data collected should be sorted into weight classes of: 0.0-0.49, 0.50-0.99, 1.0-1.49, 1.5-1.99, etc. See graph below.

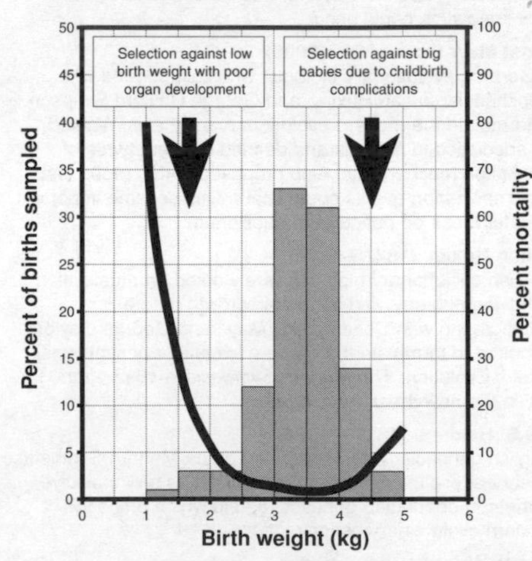

***SAMPLE DATA**: Use these data if students are unable to collect from local sources*

3.740	3.830	3.530	3.095	3.630
1.560	3.910	4.180	3.570	2.660
3.150	3.400	3.380	2.660	3.375
3.840	3.630	3.810	2.640	3.955
2.980	3.350	3.780	3.260	4.510
3.800	4.170	4.400	3.770	3.400

3.825	3.130	3.400	3.260	4.100
3.220	3.135	3.090	3.830	3.970
3.840	4.710	4.050	4.560	3.350
3.380	3.690	1.495	3.260	3.430
3.510	3.230	3.570	3.620	3.260
3.315	3.230	3.790	2.620	3.030
3.350	3.970	3.915	2.040	4.050
3.105	3.790	3.060	2.770	3.400
1.950	3.800	2.390	2.860	4.110
1.970	3.800	4.490	2.640	3.550
4.050	4.220	2.860	4.060	3.740
4.082	3.000	3.230	2.800	4.050
4.300	3.030	3.160	3.300	2.350
3.970	2.980	3.550	3.070	2.715

1. Normal distribution (bell-shaped curve), probably with a skew to the left.
2. 3.5 kg (taken from the table: only 2% mortality)
3. Good correlation. Lowest frequencies of surviving birth weights correspond to birth weights of highest mortality.
4. Selection pressures operate at extremes of the range: Premature babies have reduced survival because their body systems are not fully developed; large babies present problems with delivery as the birth canal can only accommodate babies up to a certain size. **Note**: Very large babies can occur as the result of gestational diabetes. Before adequate medical intervention, this often led to the death of the mother and/or the baby.
5. Medical intervention can now allow babies that are very premature to survive (babies as small as 1.5 kg have a good chance of survival today, but this has not historically been the case). Caesarean deliveries have also allowed larger babies to be born. **Note**: This technology is available to wealthy societies thereby reducing the effect of this selection pressure. Developing countries still experience this selection pressure.

The Evolution of Darwin's Finches (page 333)

1. Main factors contributing to adaptive radiation: Absence of competitors on the Galapagos and, partly as a consequence of this, a wide diversity of niches available for exploitation. Radiation such as this also requires a relatively unspecialized founding species with a certain amount of genetic plasticity.
2. (a) **Character displacement** is the phenomenon in which closely related species become more different (in morphology or behavior) in areas where their ranges overlap. This can be seen in the bill sizes of a number of finch species; bill sizes within a species show greater variability in the absence of potential competitors. When two finch species are sympatric, bill sizes diverge but variability within each species is minimal.
   (b) The character displacement observed in Galapagos finches provides evidence for the likely role of both allopatric and sympatric events being important in their evolution from a common ancestor. When populations are sympatric (e.g. *G. fuliginosa* and *G. fortis*) the bill sizes of the species do not overlap. Where these species exist separately on different islands (as allopatric species), intraspecific variation in bill size is greater and the bills sizes of the two species overlap over a more intermediate range. Such evidence indicates that competition in sympatric populations can lead (and had led) to character displacement to enable exploitation of different feeding niches. These differences are not observed in allopatric populations where, in the absence of competition, the feeding niche is broader and the bill size more variable.
3. (a) Evidence for **phenotypic plasticity** in Galapagos finches comes from the morphological variability (e.g. in bill size) expressed in a single species depending on whether they exist as an allopatric population or in sympatry with a closely related species. Such plasticity has its origins in the very generalized grassquit ancestor from which all thirteen species derive. (DNA analyses have confirmed all thirteen species of birds have evolved from a flock of about thirty birds arriving to the islands a million years ago).
   (b) The original finch ancestor must have been a generalist rather than a specialist, and they would have had to be able to adapt to drought, floods, and a volcanic environment in the early years of inhabitance.

Insecticide Resistance (page 335)

1. (a) and (b) any two in any order:
   - Insect populations tend to reproduce very quickly (generation times are short) and so the chances of a mutation conferring resistance (1) arising and (2) being passed on are greater.
   - Partial resistance can arise at several levels (e.g. behavioral, mechanical, biochemical) and through sexual reproduction in a given selective environment, offspring can require a sufficient number of mechanisms to develop full resistance.
   - Applications of insecticide do not always reach their intended target in the correct dosage and so do not kill 100% of the population, allowing surviving individuals to reproduce and pass on even slight resistances.
2. The insecticide application may not kill off the entire insect population. Those that survive will pass on any resistance they have to the next generation. Periodic insecticide applications act as a selection agent by allowing only the most resistant insects to survive and pass on their genes.
3. Resistance to synthetic insecticides has a number of implications. In order to maintain the kill rate of the insecticide, farmers often increase the amount/potency/toxicity of insecticide applied. This leaves more residue on the plant and so is potentially more dangerous to humans. Crops may have to be withheld longer before going to market and so cause loss of income to the farmer. Increasing resistance among insect vectors of disease is also a problem for human populations (e.g. resistance in mosquito vector for malaria), reducing the options for controlling the spread of disease through vulnerable human populations.

The Evolution of Antibiotic Resistance (page 336)

1. Antibiotic resistance refers to the resistance bacteria show to antibiotics that would normally inhibit growth. In other words, they no longer show a reduction in growth response in the presence of the antibiotic.

2. (a) Antibiotic resistance arises in a bacterial population as result of mutation. Some bacteria can also acquire these changes in DNA (conferring resistance) by transfer of genes between bacteria by conjugation (horizontal evolution).
   (b) Resistance can become widespread as a result of (1) transfer of genetic material between bacteria (horizontal evolution) or by (2) increasing resistance with each generation a result of natural selection processes (vertical evolution). In the latter case, the antibiotic provides the environment in which selection for resistance can take place. It is exacerbated by overuse and misuse of antibiotics.

3. Historically, tuberculosis was effectively treated with antibiotics, but complacency over its control has lead to increasing multiple drug resistance in the *Mtb* population and a resurgence in the number of TB cases. This has huge implications for public health because more people live with (resistant forms of) the disease and spread it to more people as a result. E - In addition, the costs associated with treating TB are now also much higher. Increasing resistance increases the costs and lowers the efficacy of treating the disease.

Gene Pools and Evolution (page 337)

1. & 2. **Note**: Do not include the beetle about to enter Deme 1 (aa) but do include the beetle about to leave Deme 1 (Aa). For the purpose of this exercise, assume that the individual with the mutation **A'A** in Deme 1 is a normal **AA** combination.

Deme 1: 22 beetles Deme 2: 19 beetles

Deme 1		Number counted	%
Allele types	*A*	26	59.1
	a	18	40.9
Allele combinations	*AA*	8	36.4
	Aa	10	45.4
	aa	4	18.2

Deme 2		Number counted	%
Allele types	*A*	13	34.2
	a	25	65.8
Allele combinations	*AA*	1	5.3
	Aa	11	57.9
	aa	7	36.8

3. (a) **Population size**: Large population acts as a 'buffer' for random, directional changes in allele frequencies. A small population can exhibit changes in allele frequencies because of random loss of alleles (failure of an individual to contribute young to the next generation).
   (b) **Mate selection**: Random mating occurs in many animals and most plants. With 'mate selection', there is no random meeting of gametes, and certain combinations come together at a higher frequency than would occur by chance alone. This will alter the frequency of alleles in subsequent generations.
   (c) **Gene flow between populations**: Immigration (incoming) and emigration (outgoing) has the effect of adding or taking away alleles from a population that can change allele frequencies. In some cases, two-way movements may cancel, with no net effect.
   (d) **Mutations**: A source of new alleles. Most mutations are harmful, confer poor fitness, and will be lost from the gene pool over a few generations. Some may be neutral, conferring no advantage over organisms with different alleles. Occasionally, mutations may confer improved fitness and will increase in frequency with each generation, at the expense of other alleles.
   (e) **Natural selection**: Selection pressures will affect some allele types more than others, causing allele frequencies to change with each generation.

4. (a) Increase genetic variation: Gene flow (migration), large population size, mutation.
   (b) Decrease genetic variation: Natural selection, non-random mating (mate selection), genetic drift.

Changes in a Gene Pool (page 341)

1. The presence of two recessive alleles means no pigment is produced. The dominant allele produces pigment. A single dominant allele produces moderate amounts of pigment to make the beetle dark. Two dominant alleles produce extra pigment to make black.

2. This exercise demonstrates how allele frequencies change as two different kinds of event take place:

Phase 1: Initial gene pool
This is the gene pool before any of the events take place:

	A	a	AA	Aa	aa
No.	27	23	7	13	5
%	54	46	28	52	20

Phase 2: Natural selection
The population is now reduced by 2 to 23. The removal of two homozygous recessive individuals has altered the allele combination frequencies (some rounding errors occur).

	A	a	AA	Aa	aa
No.	27	19	7	13	3
%	58.7	41.3	30.4	56.5	13.0

Phase 3: Immigration / emigration
The addition of dominant alleles and the loss of recessive alleles makes further changes to the allele frequencies.

	A	a	AA	Aa	aa
No.	29	17	8	13	2
%	63	37	34.8	56.5	8.7

Sexual Selection (page 342)

1. **Intrasexual selection** involves competition within one sex (usually males) with the winner gaining access to the opposite sex. Intersexual selection (or **mate selection**) also involves competition (usually involving displays) between members of one sex (usually males) for the opposite sex (usually the females) but it is the opposite sex who chooses their mate.

2. Sexual selection results in marked **sexual dimorphism** because the competitive gender (usually males) have to advertise their superiority as a mate to rivals and potential suitors. They do this by means of elaborate ornamentation (e.g. antlers, plumage) and behavior (ritualized fighting, stereotyped courtship displays). The development of these characteristics leads to greater divergence in appearance between the sexes.

Population Genetics Calculations (page 343)

The answers to the panels provided in the manual are shown below as working. Alternatively, these calculations are quickly done using a spreadsheet.

1. **Working**: q= 0.1, p= 0.9, q^2= 0.01, p^2= 0.81, 2pq= 0.18
   Proportion of black offspring = 2pq + p^2 x 100% = 99%;
   Proportion of gray offspring = q^2 x 100% = 1%
2. **Working**: q= 0.3, p= 0.7, q^2= 0.09, p^2= 0.49, 2pq= 0.42
   (a) Frequency of tall (dominant) gene (allele): 70%
   (b) 42% heterozygous; 42% of 400 = **168**
3. **Working**: q= 0.6, p= 0.4, q^2= 0.36, p^2= 0.16, 2pq= 0.48
   (a) 40% dominant allele
   (b) 48% heterozygous; 48% of 1000 = **480**.
4. **Working**: q= 0.2, p= 0.8, q^2= 0.04, p^2= 0.64, 2pq= 0.32
   (a) 32% heterozygous (carriers)
   (b) 80% dominant allele
5. **Working**: q= 0.5, p= 0.5, q^2= 0.25, p^2= 0.25, 2pq= 0.5
   Proportion of population that becomes white = 25%
6. **Working**: q= 0.8, p= 0.2, q^2= 0.64, p^2= 0.04, 2pq =0.32
   (a) 80% (c) 36% (e) 96%
   (b) 32% (d) 4%
7. **Working**: q= 0.1, p= 0.9, q^2= 0.01, p^2= 0.81, 2pq= 0.18
   Proportion of people expected to be albino (i.e. proportion that are homozygous recessive) = 1%

Analysis of a Squirrel Gene Pool (page 345)

1. Graph of population changes:

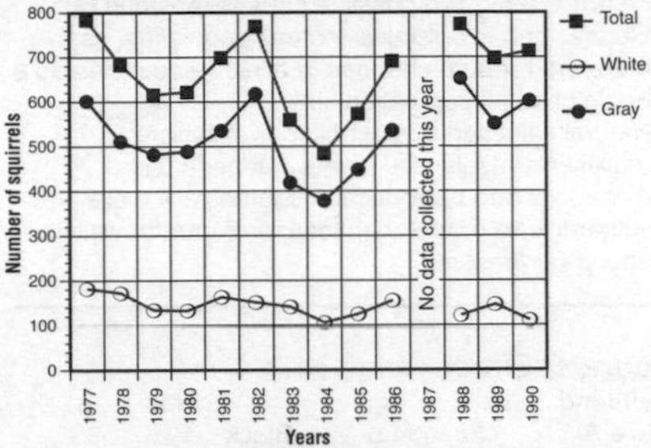

(a) 784 to 484 = 61% fluctuation
(b) Total population numbers exhibit an oscillation with a period of 5-6 years (2 cycles shown). Fluctuations occur in both gray and albino populations.

2. Graph of genotype changes:

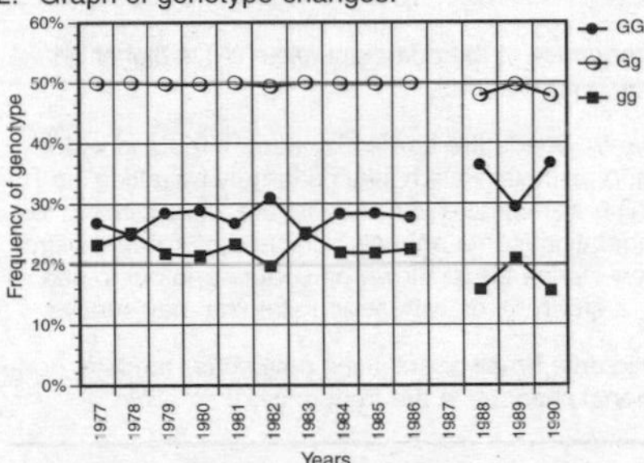

(a) GG genotype: Relatively constant frequency until the last 3-4 years, which show an increase. Possibly an increase over the total sampling period.
(b) Gg genotype: Uniform frequency.
(c) gg genotype: Relatively constant frequency until the last 3-4 years which exhibit a decline. Possibly a decrease over the total sampling period.

3. Graph of allele changes:

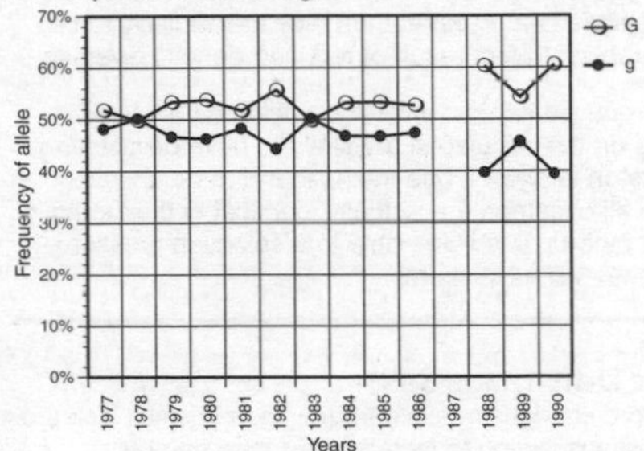

(a) Frequency of G: Increases in the last 3-4 years.
(b) Frequency of g: Decreases in the last 3-4 years.

4. (a) The *frequency of alleles* graph (to a lesser extent the *frequency of genotypes* graph)
   (b) Changes in allele frequencies in a population provide the best indication of significant evolutionary changes occurring. These cannot be deduced

simply from changes in numbers or genotypes.

5. There are at least two possible causes (any one of):
   - Genetic drift in a relatively small population, i.e. there are random changes in allele frequencies as a result of small population size.
   - Natural selection against albinos. Albinism represents a selective disadvantage in terms of survival and reproduction (albinos are more vulnerable to predators because of greater visibility and lower fitness).

The Founder Effect (page 347)

1.

Mainland	Nos	%		Nos	%
Allele A	48	54.5	Black	11	25
Allele a	40	45.5	Dark	26	59
Total	88	100	Pale	7	16
Island	**No**	**%**		**No**	**%**
Allele A	12	75	Black	4	50
Allele a	4	25	Dark	4	50
Total	16	100	Pale	0	0

2. The frequency of the dominant allele (A) is higher on the island population.

3. (a) Plants: Seeds are carried by wind, birds and water.
   (b) Land animals: Reach islands largely by rafting', in which animals are carried offshore while clinging to vegetation; some animals survive better than others.
   (c) Non-marine birds: Blown off course and out to sea by a storm. Birds with strong stamina may survive.

4. Genetic drift: Small populations may suffer random, non-directional changes in the frequency of an allele.

Population Bottlenecks (page 348)

1. A sudden decrease in the size of a population can result in a corresponding reduction in genetic variation. This means the population has limited 'genetic resources' to cope with the selection pressures imposed on it. In particular, it is seen as reduced reproductive success and greater sensitivity to disease.

2. Poor genetic diversity means that if one individual is susceptible to a disease, then they are all likely to be vulnerable; a direct result of reduced genetic diversity.

3. With reduced genetic diversity, selection pressures acting on the population are likely to have devastating effects on survival if one trait is found to be unsuited. Since all cheetahs are virtually identical in their traits, if one individual is vulnerable to a selection pressure, then they will all succumb.

Genetic Drift (page 349)

1. Random changes in allele frequencies in small isolated populations, owing to factors other than natural selection. Not all individuals, for various reasons, will be able to contribute their genes to the next generation.

2. Genetic drift reduces the amount of genetic variation in very small populations. Alleles may become eliminated altogether (0%) or become fixed (100%) as the only allele present in the gene pool for a particular gene.

3. Any endangered species (or subspecies) with small numbers of individuals remaining, e.g. Puffin, monk seal, European lynx, otter, European beaver, hawksbill turtle, oryx, Siberian tiger (subspecies) (pop. ~500), Chinese tiger (subspecies) (pop. ~250), Sumatran tiger (subspecies) (pop. ~500), humpback whale, gray whale, blue whale, gharyal, Orinoco crocodile (<100), kakapo (New Zealand native ground parrot) (~100).

Isolation and Species Formation (page 350)

1. Isolating mechanisms protect the gene pool from the diluting and potentially adverse effects of introduced genes. Species are finely tuned to their niche; foreign genes will usually reduce fitness.

2. (a) Geographical isolation physically separates populations (and gene pools) but, if reintroduced, the two populations could potentially interbreed, i.e. reproductive isolation may not have occurred.
   (b) Geographical isolation enables populations to diverge in response to different selection pressures and (potentially) develop reproductive isolating mechanisms. Reproductive isolation won't generally occur in a populations in which there is gene flow (unless by special events such as polyploidy).

3. Geographical isolation physically separates populations (gene pools) so there is no gene flow between them. Ecological isolation arises as a result of different preferences in habitat or behavior even though the populations occupy the same geographical area.

Reproductive Isolation (page 351)

1. (a) Postzygotic: hybrid breakdown
   (b) Prezygotic: structural
   (c) Prezygotic: temporal
   (d) Postzygotic: hybrid inviability

2. They are a secondary backup if the first isolating mechanism fails. The majority of species do not interbreed because of prezygotic mechanisms. Postzygotic mechanisms are generally rarer events.

Allopatric Speciation (page 353)

1. Interspecific competition, intraspecific competition (this is the strongest). New habitat becomes available (loss of geographical barrier).

2. (a) Plants move by dispersal of their seeds (water borne, carried vast distance by winds, animals carry them in their fur/feathers or in their gut to be deposited with their feces).
   (b) Gene flow between the parent population and dispersing populations is regular. Gene flow refers to the movement of genes, via gametes, from one population to another.

3. Orogeny or plate tectonics (continental drift).

4. Ice ages, glacials, or interglacials.

5. Mountains form. Rivers form or change course. Deserts. Isthmus or land bridge forms (land separating two seas, e.g. Central America). Ice sheets advance. Sea or ocean develops (due to continental drift).

6. Physical barriers prevent gene flow between separated populations.

7. (a) Selection pressures on gene pools (any four): Various climatic factors (e.g. temperature, pH, salinity, wind exposure), competition, predation, disease and parasitism, quality of the food resource.
   (b) Some individuals will have allele combinations (and therefore a phenotype) that better suits the unique set of selection pressures at a given location. Over a period of time (generations) certain alleles for a gene will become more common in the gene pool, at the expense of other less suited alleles.

8. Reproductive isolating mechanisms: (see pages 119-120 of the workbook for full descriptions).
   (a) **Prezygotic** (prevent fertilization from taking place): spatial, ecological, temporal, behavioral, structural, gamete mortality.
   (b) **Postzygotic** (fertilization occurs, but the result is unsuccessful, to varying degrees): zygote mortality, reduced fertility, hybrid breakdown.

9. **Allopatry** refers to the situation where populations within the species have separate areas of geographical range. **Sympatry** refers to the situation where populations within the species inhabit the same or overlapping geographical areas.

Sympatric Speciation (page 355)

1. A speciation event that occurs without prior geographical separation. The two species are separated by some other means such as niche differentiation, or may be separated by a spontaneous chromosomal change (polyploidy).
2. Polyploidy creates extra sets of chromosomes for an individual that make it impossible for it to reproduce with members of its 'parent' population. Hybrids may form but they will be sterile.
3. Modern wheat (see the activity *Breeding Modern Wheat*), many modern fruit and vegetable varieties, e.g. kiwifruit, fuji apple, banana, boysenberry, strawberry.
4. If two groups within a species population have slightly different niches (e.g. different habitat preferences), then they will not come into contact for mating.

Stages in Species Development (page 356)

1. Gene flow becomes reduced and the two populations evolve in different directions (until gene flow ceases).
2. Considerable geographical barriers would have prevented much gene flow. The large distances between the sparsely populated communities would have made contact a rare event, even for nomadic people. It may have been seen as desirable to introduce 'new blood' into a group when meeting others and women may have been exchanged. More 'hostile gene flow' may have occurred if sexual intercourse was forced upon women during attacks between warring peoples.
3. (a) An increasing number of hybrids leads to genetic swamping of the non-hybrid species. As non-hybrid breeding partners become rarer, this process accelerates, until few non-hybrids remain.
   (b) Some ducks may still be deterred by mating behavior that does not match exactly that required by their own species.

Selective Breeding in Animals (page 357)

1. **Inbreeding** involves breeding between close relatives, and if practiced over a number of generations lead to increased homozygosity in a population. It is used by animal breeders to 'fix' desirable traits into a population, but an increase in the frequency of recessive, deleterious traits in homozygous form in a population can reduce the health and fitness and of a population and lower fertility levels. **Out-crossing** involves introducing new (unrelated) genetic material into a breeding line. It is used to increase the genetic diversity, and is used in line-breeding to restore vigor and fertility to a breeding line.
2. Assisted reproductive technologies, such as artificial insemination, cryopreservation, embryo transfer and *in vitro* fertilisation, are used routinely to produce large numbers of offspring with desirable traits (e.g. high growth rates or superior wool production). These techniques allow the desirable traits to be fixed more quickly into the population than would be possible from traditional selective breeding techniques.
3. Selective breeding over time has removed the genes that coded for the more aggressive characteristics from the gene pool of domestic dogs while retaining genes for more useful characteristics such as patience and obedience. This has been achieved by selectively breeding dogs that displayed these desirable characteristics. The development of breeds has caused specific genotypes to appear in the gene pool that further enhance these or other desirable characteristics.
4. Wolves are highly social animals that live in often large groups (packs). Hierarchy and appeasement behavior bind the group and reduce aggression.
5. Dog breed traits selected for:
   (a) Hunting large game: Good sense of smell, strong bite and strong neck muscles, fearless, aggressive.
   (b) Game fowl hunting: Excellent sense of smell (detection), good eyesight, understanding of need to 'hold', 'point', and retrieve.
   (c) Stock control: Must not regard stock as prey (low aggression to stock), obedience (good at taking instructions from farmer, ability to anticipate the behavior of stock animals and respond to stock movements, bark and use body language to direct stock movement, protect stock from predators.
   (d) Family pet: Low level of aggression, playful attributes, friendly disposition.
   (e) Guard dog: Aggressive behavior to strangers, excellent hearing and sense of smell, alert to the arrival of intruders, respond by vigorous barking.

Selective Breeding in Plants (page 359)

1. (a) Cauliflower: flowers
   (b) Kale: leaf
   (c) Broccoli: inflorescence
   (d) Brussels sprout: lateral buds
   (e) Cabbage: apical (terminal) bud

(f) Kohlrabi: stem (swollen)

2. If allowed to flower, all six can cross-pollinate.

3. Unwanted plant species in our gardens (weeds) are selected against by a number of control methods. Physical weeding by hand or digging implement will favor those plants that have tough roots (e.g. dock) or propagation methods that are stimulated by weeding (e.g. oxalis). Heavy use of sprays will foster the development of herbicide resistance.

4. Broccoli is an inflorescence, therefore breeders would have selected for plants exhibiting clumps of many small flowers. They would also have selected for plants producing thick fleshy flower buds. These two main characteristics, selected together over generations of plants, eventually produced modern broccoli.

5. Desirable characteristics of apple trees (any of): development of sweet fruit, with crisp, juicy flesh, fruit that remains on the tree until picked (ripe), trees that grow to a uniform shape and size, trees that produce fruit of a uniform shape and size, etc.

6. (a) Selective breeding for specific traits generally reduces genetic diversity by increasing homozygosity in the offspring. When selection in focussed on specific traits, other phenotypes (therefore genotypes) are rejected and their genes are lost from the gene pool. This is particularly the case when the genes for desirable traits are associated, e.g. a genotype for heavy fruiting might also be associated (e.g. through linkage) to lower seed production. Selection for one trait will then also select for another.
   (b) Retention of genetic diversity is particularly important in crop plants because it provides a pool of genes from which to improve strains and guard against loss of adaptability in crops. In terms of food security, it is dangerous to rely on only a restricted number of strains for most of our food. A good example is the Irish potato famine where potatoes were the main food crop and farmers relied almost exclusively on one high yielding potato variety. When this variety proved vulnerable to blight, most of the country's crop was lost and there was a huge famine. The country lost food security by relying on one variety and by not having a readily available store of diversity on which to draw.

7. Cultivated American cotton would have originated from the interspecific hybridisation of Old World cotton and wild American cotton.

8. Cavendish bananas do not produce seed and therefore must be reproduced by asexual propagation. As a result, all Cavendish banana plants are genetically identical and all will be vulnerability to the Panama disease strain. Moreover, because the plant is sterile, it is not possible to breed into it any more genetic diversity and it will not be able to naturally produce resistance to the disease within the population (except perhaps in the unlikely event of a favorable mutation).

9. Wild plants and ancient breeds possess alleles that may have been lost from inbred lines. The retention of these ancient cultivars provides a gene bank and a buffer of genetic diversity which can be used to improve the inbred cultivars in the future.

Breeding Modern Wheat (page 361)

1. A hybrid of inbred lines increases heterozygosity in the offspring. This is associated with a phenotypic response called hybrid vigor, characterized by greater adaptability, survival, growth, and fertility.

2. (a) and (b) any of the following:
   - Seeds that do not scatter but remain on the plant until harvest.
   - Large seed heads containing large amounts of gluten.
   - Plants that grow in uniform shape and size.
   - Plants with natural high resistance to disease.
   - Plants tolerant of a range of growing conditions.

KEY TERMS: Crossword (page 362)

Answers Across

1. Polymorphism
2. Species
4. Frequency
5. Gene pool
7. Gene flow
11. Geographic
13. Microevolution
14. Deme
15. Ring species
16. Sympatric

Answers Down

1. Prezygotic
3. Evolution
6. Directional
8. New synthesis
9. Allopatric
10. Hybrid
12. Fitness

Patterns of Evolution (page 363)

1. (a) Sequential
   (b) Divergent
   (c) From common ancestor D: W, B, P, and H
   From common ancestor B: P and H
   (d) The trait was first developed in species B, then passed on to P and H.

2. (a) **Divergence** refers to the evolution of two (or more) species from a common ancestor (it is also called branching evolution). **Adaptive radiation** is a form of divergence where many species evolve from a common ancestor to occupy a range of new niches.
   (b) **Divergent** evolution involves branching of two or more species from a common ancestor. In contrast, **sequential** evolution involves evolutionary change in a single species over time without branching. In sequential evolution, one species accumulates changes gradually so that, over time, it becomes different enough to be classified as a new species.

Convergent Evolution (page 365)

1. (a) Streamlined body shape to reduce drag in the viscous, aquatic environment.
   (b) Forelimbs developed into flippers for efficient propulsion through the water, as well as control and maneuverability.

2. (a) Need for speed to catch fish and avoid predators.
   (b) Need for maneuverability (quick turns) for the same reasons as above.

3. The similarities are due to convergence rather than common ancestry. The taxonomic groups (of species)

tend to evolve from common ancestors locally and undergo adaptive radiation.

4. (b) **Flying phalanger / flying squirrel**
   Adaptations: Both have similar body size and shape. Skin stretched between forelimbs and hind limbs to enable gliding between trees. Both eat insects and some plants.
   Selection pressures: Predator avoidance on the ground as well as an energy efficient means of travelling between trees in search of food resources.

   (c) **Marsupial mole / mole**
   Adaptations: Reduced vision, streamlined body shape, powerful forelimbs with effective shovel-like claws for digging through soft soil. Both eat insects.
   Selection pressures: Digging into earth and moving through small borrows underground.

   (d) **Marsupial mouse / mouse**
   Adaptations: Small size, agile climbers inhabiting low shrubs in dense ground cover. Active at night.
   Selection pressures: Predator avoidance (nocturnal habit) and ability to exploit food resources on bushes capable of taking only a little weight.

   (e) **Tasmanian wolf (tiger) / wolf**
   Adaptations: Strong skull and associated musculature, with canine and carnassial teeth adapted for shearing and tearing meat. The limb bones are long and slender, suitable for running.
   Selection pressures: Catching, killing, and dismembering prey. Ability to pursue prey over sustained distances with rapid bursts of speed for prey capture.

   (f) **Bandicoot / rabbit**
   Adaptations: While the rabbit is primarily a vegetarian, the group of marsupials known as long-eared bandicoots are varied in their diet. Some are plant eaters while others eat insects. Both have long ears for hearing potential danger signals, and long hind legs adapted for a hopping form of locomotion.
   Selection pressures: Predator avoidance: detecting the approach of danger and then escaping.

Coevolution (page 367)

1. **Coevolution** is a change in the genetic composition of one species (or group) in response to a genetic change in another, i.e. it is reciprocal evolutionary change in interacting species. Examples include predator-prey and parasite-host relationships and mutualistic relationships such as those between plants and their pollinators.
2. Flowers advertising the presence of nectar and pollen have evolved to attract insect pollinators. This includes color, scent, shape and arrangement. Even the time of day of flower opening can coincide with the activity of a desirable pollinator. Some flowers reflect UV and act as nectar guides to guide the pollinator to the nectary.
3. Coevolutionary events, in which adaptive changes in one species follow adaptive changes in another, lead to an increase in biodiversity by creating specific intimate ecological relationships from initially casual and broader interactions. Increasing specialization in this way partitions broad niches so that a larger number of specialized niches become available. Coevolution between flowering plants and beetles illustrates this phenomenon. Beetles are a very ancient group of insects, now with thousands of modern species, and a vast number of flowering plant species depend on particular beetle species for pollination. In many cases, the relationship is one of mutual interdependence.
4. Competition between species, although it can lead to competitive exclusion of one species at the expense of another, may also result in niche differentiation and character displacement (morphological divergence of species when they occur in the same range). Niche differentiation provides the opportunity for specialization and the development of close, even interdependent, ecological relationships between different species. For example, competition for pollinators may lead one plant species to make adaptive changes to enhance success with a particular pollinator. The pollinator responds with adaptive changes of its own to enhance the efficiency of the relationship and so on. The two species then coevolve and occupy a plant/pollinator niche previously not in existence.
5. The analogy of the coevolutionary "arms race" does not hold for a parasite and its host because as soon as the parasite evolves further in its tenuous relationship with its host, there is a good chance that the host would be killed. A successful parasite is one that maintains a living host.

Pollination Syndromes (page 369)

1. (a) For example, any one of the following:
   Beetles have coevolved to pollinate ancient plant groups; their hard, smooth bodies and good sense of smell are adapted to pollinate large, easy access flowers with strong fruity odors.
   Nectar-feeding flies sense nectar with their feet and have tubular mouthparts suited to feeding from easy access, odorless flowers.
   Moths are active at night and have a good sense of smell and the plants they pollinate thus tend to have heavily scented flowers that open at night. The flowers pollinated by moths also tend to provide a landing platform, which many moths require. Moths have long, narrow tongues, and the flowers of the plants they pollinate house their nectar in narrow, deep tubes.
   Carrion flies are attracted by heat, odor, and the color of dung or carrion, and the plants they pollinate have features appropriate to attracting them (e.g. they produce foul odors or heat). Carrion flies do not require nectar or pollen and so none is provided by the plant.

   (b) For example, any one of the following:
   Birds, which are largely diurnal with high energy needs and good color vision, pollinate flowers that open during the day, and are large and damage resistant, not particularly fragrant, brightly colored (often red), and produce copious nectar.
   Bats, which are nocturnal, have high energy needs and a good sense of smell, but are color blind. Consequently the flowers they pollinate open at night, are strong smelling but dully colored, produce plentiful nectar and pollen, and have pendulous blossom or blossom on the tree trunk (since bats cannot fly in dense foliage).
   Non-bat mammals, e.g. opossums and rats, that act as pollinators tend to be relatively large with a good sense of smell and high energy requirements.

The plants they pollinate are thus robust and damage resistant, dull colored but odorous, and produce copious sugar-rich nectar.

2. Knowledge of what animal pollinates a certain plant would allow a prediction to be made about the type of pollination of a similar plant.

The Rate of Evolutionary Change (page 370)

1. (a) **Punctuated equilibrium**: Rapid changes in the environment i.e. sudden habitat changes or the introduction/removal of other species.
   (b) **Gradualism**: Gradual but steady changes in the environment allowing species to develop adaptive features slowly.
2. Punctuated equilibrium.
3. (a) Any two of: Sharks, horseshoe crabs, coelacanth.
   (b) Stasis.
   (c) There is no selection pressure for change; they are superbly well adapted (for their niche) as they are. There is suggestion that such morphologically static species may track environmental conditions through space and time, and so be subjected to conservative selective pressures throughout their history.

Adaptive Radiation in Ratites (page 371)

1. (a) Flightlessness (wings very reduced in size), flat breastbone, primitive palate.
   (b) Primitive palate.
2. (a) Anatomical change: Reduction in wing size. Selection pressure: Advantages gained by putting energy into developing other parts of the body, rather than into a structure (wing) that is not used.
   (b) Anatomical change: Larger, stronger legs. Selection pressure: A need to improve the strength of the legs for locomotion after losing the ability to fly. Legs would have been used both for escape and as weapons to repel predators.
3. Gondwana.
4. (a) Ostrich (from Africa)
   (b) It has the closest distribution to Madagascar.
   (c) Divergence of the elephantbird:

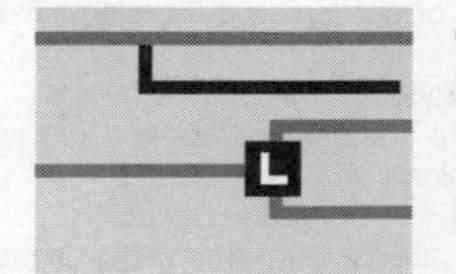

Ostrich
Elephantbird
Rhea 1
Rhea 2

5. (a) Other flightless birds: penguins, flightless cormorant, takahe (a rail), great auk (extinct), dodo (extinct), kakapo (glides but does not fly).
   (b) They have evolved flightlessness independently of the ratites, arising from different common ancestors. Flightlessness in birds is restricted to islands with few, if any, large terrestrial predators.
6. The moa species arose as they diversified to occupy different niches. A range of sizes in the eleven species meant that they exploited food resources at different heights in the vegetation. Different habitat preferences, ranging from alpine to coastal, further separated the niches of these herbivorous birds. Kiwis are highly specialised nocturnal feeders, adapted to exploit a very narrow food resource.
7. The moas were already established in New Zealand before it separated from the Gondwana super continent. Once the separation was underway, a later arrival of the kiwi ancestors occurred. There is still speculation as to whether these kiwi ancestors had already lost the power of flight. If they had limited flying ability, it has been suggested that they may have island-hopped via a chain of islands between Australia and New Zealand (what is now the Lord Howe Island Rise). If this island chain existed, it may have been lost since, as a result of plate tectonic activity in the Tasman Sea (information provided by Alan Cooper – *pers. comm.*)
8. (a) Common ancestor: F
   (b) Common ancestor: C

Adaptive Radiation in Mammals (page 373)

1. **Adaptive radiation** (the evolution, from a common ancestor, of a diversity of related species occupying many different niches) usually involves a rapid increase in diversity and morphological change within a lineage. The mammals share a distant common ancestor 195 mya; long before their main period of adaptive radiation. For a long period of their evolutionary history, there was very little diversification, with the exception of the early divergence of the marsupials. The diversity of form and function now represented by the placental orders arose comparatively quickly as a result of diversification into niches vacated after the demise of the dinosaurs.
2. Ancestral forms, common ancestors
3. **B**: A divergence occurred (one group of early mammals split into two evolutionary lines: the marsupials and the placentals).
4. (a) **D**: Were ancestral forms to many other orders of mammal.
   (b) **D**: Became extinct (no examples surviving now).
5. Rodents and odd-toed ungulates (widest gray shape which indicates the largest number of species).
6. **A**: Evolutionary lines that became extinct (dead ends).
7. Paleocene
8. (a) and (b) in any order:
   – They have body hair (although there is some evidence to suggest that flying dinosaurs may also have had body hair).
   – They provide milk from mammary glands to nourish offspring.

   Note that these are shared, derived characters, and therefore distinguish the class (as opposed to shared, ancestral characters, such as a vertebral column).
9. Principal reproductive features:
   (a) Monotremes: Lay eggs.
   (b) Marsupials: Give birth after a short gestation. The poorly developed (essentially embryonic) young climb into a pouch and attach to a nipple.

(c) Placentals: Fetus is nourished by a placenta (blood supply from the uterus) and born in a relatively more advanced state.

10.
1	Probiscidea	10	Artiodactyla
2	Rodentia	11	Dermoptera
3	Xenarthra	12	Sirenia
4	Chiroptera	13	Macroscelidae
5	Perissodactyla	14	Insectivora
6	Carnivora	15	Tubulidentata
7	Pinnipedia	16	Hyracoidea
8	Lagomorhpa	17	Pholidota
9	Primates	18	Cetacea

Note: Some taxonomists vary in their interpretation of mammalian classification, e.g. Pinnipedia are often included with the Carnivora; older classifications place the elephant shrews in the Insectivora (their closest relatives are now considered to be the lagomorphs).

11. Adaptations of the mammalian groups (a)-(c) any of:

Order: Probiscidea: elephants.
Adaptive features: Large trunk (fusion of nose and upper lip) and large body size.
Advantage: Enables browsing in a variety of habitats (African wooded savanna to tropical rainforests of Asia).

Order: Rodentia: rats, mice, squirrels, beavers, porcupines.
Adaptive features: Single pair of chisel-like incisors that grow throughout life and are used for gnawing.
Advantage: Predominantly eaters of seeds and tough vegetable matter, their teeth are well suited for the task.

Order: Xenarthra: anteaters, armadillos, sloths. Comprises a group of mammals that radiated in South America 65-2 million years ago. Four main lines evolved from this ancient stock, 3 survive today:
Adaptive features and advantages:
1. Armadillos: armored terrestrial grazing herbivores.
2. Sloths : tree-dwelling non-armored group specializing in browsing.
3. Anteaters: extremely well adapted for feeding on ants and termites.

Completely toothless or possessing just primitive molars. Anteaters have powerful forelimbs with large sharp claws, and a long nose equipped with a long sticky tongue. Sloths rarely descend to the ground and have limbs adapted for hanging, with claws adapted as hooks. Armadillos are covered with a flexible horny shield backed by bone and many can curl up into a defensive ball.

Order: Chiroptera: bats.
Adaptive features: The forelimb has greatly lengthened bones (especially the digits) with skin stretched between, to form a wing. Echolocation is highly developed in some.
Advantage: The only true flying mammal, allows exploitation of tree resources and airborne predation. Echo location allows nocturnal hunting.

Order: Perissodactyla: horses, tapirs, rhinoceroses
Adaptive features: One of the orders of hoofed mammals, these animals are adapted to walk and run on the tips of 1 or 3 toes. Specialized grazers with guts adapted for dealing with large volumes of plant material.
Advantage: Their foot structure has allowed most of these animals to achieve high speeds in running. Digestion of plant cellulose when browsing allows exploitation of vast food resources.

Order: Carnivora: dogs, cats, bears, raccoons.
Note: Not all members of the order Carnivora are carnivores (e.g. Panda bears eat only bamboo, but scavenge meat) and many carnivores are found outside this order. In fact, most members of this order are strictly omnivores. It is their dentition that defines the group.
Adaptive features: Legs adapted for fast running. Teeth suited to crush bones and cut meat. Many have developed cooperative hunting behaviors.
Advantage: Allows pursuit of prey and dismembering of a carcass.

Order: Pinnipedia: seals, walruses, sealions. (Note: often included as a suborder of Carnivora).
Adaptive features: Adaptation to the marine environment involve streamlined bodies and limbs modified to form flippers.
Advantage: Features allow fast movement through the water which allows exploitation of the vast fish resource.

Order: Lagomorpha: rabbits, hares, pikas.
Adaptive features: Have two pairs of incisor teeth compared to the rodent's one pair. Adaptations include large ears and widely separated eyes, and in many cases hind legs for rapid hopping movement.
Advantage: These adaptations are largely directed by predator evasion.

Order: Primates: prosimians, monkeys, apes, humans.
Adaptive features: Have nails rather than claws, thumb and big toe which are usually opposable with other digits, relatively large brain and well developed eye-sight (often binocular).
Advantage: Ideally suited to tree-dwelling allowing tree resources to be exploited. (Humans later developed bipedalism from a tree dwelling habit).

Order: Artiodactyla: pigs, camels, deer, cattle, antelope.
Adaptive features: These cloven-hoofed mammals have 2 or 4 weight- bearing toes on each foot. This group has a high species diversity. Generally herbivores, although some pigs are omnivores. The hoof foot structure gives immediate traction on almost any earth surface when power is massively and suddenly applied. Complex stomachs allow fermentation of plant food.
Advantage: Adaptations are dominated by means of evading predators and processing plant food (grazing).

Order: Dermoptera: culogos (flying lemurs).
Adaptive features: Gliding membrane that stretches from the neck to the fingertips, along the sides of the body and joins the legs and tail. They are herbivores and can glide up to 135 m from tree to tree.
Advantage: Gliding enables them to move quickly from tree to tree while searching for food in the canopy.

Order: Sirenia: dugongs, manatees (sea cows).
Adaptive features: Body is streamlined with molar teeth that grow continuously to counter the abrasive effects of fresh-water plants with a high silica content. Mostly marine and living in estuaries and river systems.
Advantage: Able to exploit a food resource with little or no natural predators.

Order: Macroscelidae: elephant shrews
Adaptive features: Elongated snout, with nostrils at the tip, long hind legs, relatively large eyes and ears, nocturnal behavior if harassed by predators.
Advantage: Snout effective at probing through leaf litter in search of invertebrate prey. Other adaptations relate to rapid escape from or avoidance of predators.

Order: Insectivora: hedgehogs, shrews, moles
Adaptive features: Share a tendency (in some cases extreme specialization) toward eating insects. Small, highly mobile animals with long, narrow and often elaborate snouts with unspecialized teeth and limbs. The taxonomy of this primitive mammal group is still under much debate - a dumping ground for those species that do not share clear affinities with other groups or warrant their own order.
Advantage: Well suited for the exploitation of small terrestrial invertebrates.

Order: Tubulidentata: aardvark.
Adaptive features: Forelimbs are short, powerful digging tools that have 4 shovel-shaped claws on each foot. Their mouth is small with a long narrow tongue.
Advantage: Feeds predominantly on ants and termites. This is not an "anteater" in the strictest zoological sense, but an example of convergent evolution. Forelimbs and tongue provide excellent adaptations to exploit the rich resource of ants and termites.

Order: Hyracoidea: hyraxes
Adaptive features: Walk on the soles of their feet which are unique; they are large, rubbery-soft elastic pads kept moist by secretory glands.
Advantage: Hyraxes live among rocks or are tree climbers, so this type of foot enables a firm grip on rocks and trees, allowing exploitation of resources in a variety of herbivorous niches.

Order: Pholidota: pangolins (scaly anteaters).
Adaptive features: Body scales protect every part of the body except the underside. Some are terrestrial while others are tree-dwelling. Feeding exclusively on insects (especially ants and termites) they have powerful short limbs which are used for digging into ant hills and termite mounds. No teeth, but a sticky tongue, up to 70 cm long, is used to extract insects from their hiding places.
Advantage: Body scales afford protection from predators. Limbs and tongue allow exploitation of insects as food.

Order: Cetacea: whales, dolphins
Adaptive features: Streamlined bodies and their ability to dive to great depths for long periods of time. Baleen whales have fibrous plates in their mouths to filter out krill (shrimps).
Advantage: Large, powerful hunters and filter-feeders are able to exploit the bountiful resources of the sea. Deep diving avoids competition with surface predators.

Note: There are many examples of convergence amongst the mammals (e.g. pangolins, anteaters, aardvarks).

Geographical Distribution (page 375)

1. Once the early camels moved into Asia and northern Africa, already established competitor and predator species may have prevented the expansion of their range further south.
2. Camels were imported from northern India as beasts of burden, mainly between 1860 and 1900. After the 1920s, out-competed by trucks and trains, many were shot but some were released into the wild where they have thrived. They could not have moved into Australia without human assistance because there has never been a continuous land bridge between Australia and the Asian continent.
3. Camels died out in North America during the last Ice Age. Reasons may have included human predation or an inability to adjust to the climate change and resulting changes in the habitat.
4. A land bridge across the Bering Strait formed 1 mya. Until this time the strait provided an effective barrier to the expansion of the early camel ancestors into Asia.
5. North Africa, Middle East, Central Asia, Andes in South America. Distribution is scattered due to local extinction of populations and species in the areas in between.

Extinction (page 376)

1. (a) **Permian extinction**: 225 mya. Extinction of nearly all life on Earth including 90% of marine species. 84% of genera were lost, including many lampshells, and all trilobites and sea scorpions.
   (b) **Cretaceous extinction**: 65 mya. Extinction of more than half the marine species and many families of terrestrial plants and animals, including nearly all the dinosaurs (with the exception of ancestral birds).
   (c) **Megafaunal extinction**: 10 000 years ago; the Pleistocene extinction. Characterized by the loss of the giant mammalian 'megafauna' which proliferated through the latter part of the Tertiary period.
2. Human activity has contributed to extinctions through the destruction and pollution of habitats, accelerated climate change, and direct destruction of species (hunting for food, trade, or 'sport').
3. Extinction is a natural and necessary part of the evolutionary process. It provides opportunities for adaptive radiations, with species evolving to occupy vacant niches left behind by extinct species.

Causes of Mass Extinction (page 377)

1. (a) **Asteroid/comet impact**: An impact by a large extraterrestrial object such as a comet or asteroid would throw up a huge dust cloud, suppress photosynthesis, and cause the collapse of food chains. **Note**: The largest herbivores would initially be the most vulnerable animal species, as they have the largest food requirements and are directly dependent on plant food. Predators would then become increasingly vulnerable.
   (b) **Continental drift**: Continental drift would cause massive changes in the ocean currents (e.g. Gulf Stream), air flow patterns, and climate (and therefore the extent and type of habitats). Species unable to move or adapt would become extinct.

Note: At times in the past, continental movements have generated periods of warm tropical climate over a large geographical range. At other times, the climate has cooled so much that ice has extended over large areas.

(c) **Volcanism**: Volcanic activity would cause massive amounts of dust and poisonous gases to be released into the atmosphere, producing a screen which would reduce incoming sunlight. This in turn would cause a rapid cooling of the global climate, a fall in photosynthesis, and a change in availability and suitability of habitats. These changes could bring about mass extinctions. **Note**: Periods of volcanic activity on a massive scale (such as occurred in the Deccan Traps in India at the end of the Cretaceous) may have contributed to the extinction of the dinosaurs. Climate change was noted after the eruption of Mount Pinatubo in 1991.

2. The arrival of a new plant or animal species on a continent could displace an existing species and cause its demise. **Note**: The introduced species may be a more aggressive competitor in the same or a related niche, it may be a direct predator of the native species, or it may be a pathogen or a vector of disease.

KEY TERMS: Word Find (page 378)

```
S N H S I X T H E X T I N C T I O N R T A S M T
D I V E R G E N T E V O L U T I O N B X E O A K
E S G E Y J E X T I N C T I O N L T T I J X Q Q
B V P V D D I Z C S S F N V T U W E C C X G Y M
O K H O R G R A D U A L I S M I E E H R N G Y M
Z P Y D F V Y H C O B T C G D M P I R H P J K O
K R L E A Z X C U F U F X E O S S G U R N L P D
Z I O V R S O L O L G D E V N I L U I Y I H N T
F A G O Y M R C O E V O L U T I O N Q S Y Q F Y
K E E H O M O L O G O U S S T R U C T U R E S S
O K N X Y C L H O X G E N E S O B M N Y K G G T
P O Y M O N O P H Y L E T I C M Y C I N O P U J
N V S Q Z U I H D M W E H T Z X P E N Z D H O A
A T I X C O N V E R G E N T E V O L U T I O N X
S I E P A R A L L E L E V O L U T I O N N X B K
P U N C T U A T E D E Q U I L I B R I U M C O S
L V A D A P T I V E R A D I A T I O N O H N R A
S E Q U E N T I A L E V O L U T I O N A X A Q I
W Q I J S O H Z Y U O X S W G V E V Z V Z G O H
W P O A N A L O G O U S S T R U C T U R E V Y P
```

Notes

Notes